DIVIDENDISM

*Post-Labor Economics and the
Path to Human Liberty*

David Khoa Nguyen

GLITCH PRESS, The

First Edition: 2026 Printed in the United States of America
ISBN (Hardback): 979-8-9941862-3-7
ISBN (Paperback): 979-8-9941862-7-5
ISBN (Ebook): 979-8-9941862-5-1

To the last generation of workers, and the first generation of free humans.

CONTENTS

PREFACE: THE VIEW FROM THE METAL

I am not an economist. I do not sit in an ivory tower debating interest rates or the price of wheat. I do not have a tenure-track position at a university where the curriculum hasn't changed since the Cold War.

I am a Test & Debug Architect.

I spend my days standing in the cold aisles of data centers, listening to the deafening hum of the future being born. I work with the hardware—the NVIDIA H200s, the Blackwell B300 superclusters—that you only read about in frightening headlines. I touch the metal that is currently learning how to out-think, out-create, and out-produce the human race.

When you work at the hardware level, you don't see "politics." You don't see "ideology." You see **Physics**.

You see heat. You see voltage. You see throughput. And most importantly, you see **Velocity**.

I am writing this book because I am witnessing a terrifying discrepancy in velocity.

Inside the server room, we are moving at the speed of light. We are doubling the intelligence of the machine every six months. We are building a "Silicon God" that solves problems in seconds that would take a human lifetime. We are entering the 22nd Century.

But outside the server room? Look at the street. Look at the

Senate. Look at your bank account.

We are still running the economic software of the 19th Century.

We are still operating on the "Fordist Code" of 1914: *Human trades labor for wage; wage buys product; product funds corporation.*

This compatibility error is no longer just a glitch. It is a fatal system failure. The hardware (AI) has become too powerful for the software (The Wage System). The engine is revving to 20,000 RPMs, but the transmission is stuck in first gear.

You feel this vibration, don't you?

You feel it in the grocery store when you look at the price of beef. You feel it in the knot in your stomach on Sunday nights. You feel it when you apply for a job and realize you are competing against an algorithm that never sleeps and asks for no salary.

You are not crazy. You are sensing the **Zero Point**.

You are sensing the moment where the value of human labor crashes into the floor, while the value of assets shoots to the moon.

The economists will tell you this is a "soft landing." The politicians will tell you it is a "transitory period." They are lying to you. They are trying to patch a collapsing dam with duct tape.

I wrote this book because I am tired of the lies. I wrote this book because the solution to a broken machine isn't hope—it's engineering.

We do not need a new tax policy. We do not need a new President. We need a **System Rewrite**.

We need to uninstall the "Worker" update and install the "Shareholder" update. We need to stop fighting the machines and start charging them rent. We need to transition from an economy of Scarcity (Labor) to an economy of Abundance (Dividends).

This book is that blueprint.

It is not a polite suggestion. It is a **Mandate**.

It is a manual for the next ten years of human history. It covers the **Diagnosis** (why you are broke), the **Theory** (why it's getting worse), the **Solution** (how we fix the nation), and the **Action** (how you survive the transition).

Do not read this looking for comfort. Read this looking for weapons.

The old world is burning down. The new world is being built in the dark, on the servers I debug every day.

The robots are here to work.

We are here to live.

It is time to architect the deal.

David Khoa Nguyen

The System Architect

2026

PART I: THE DIAGNOSIS - THE BROKEN ENGINE

CHAPTER 1: THE FORD PARADOX & THE SUICIDE OF CAPITAL

Section 1: The Trap of 1914

Highland Park, Michigan. January, 1914.

Close your eyes. Listen.

It doesn't sound like innovation. It sounds like a catastrophe. It is a deafening, rhythmic grinding of metal against metal, a cacophony so loud that men have to scream into each other's ears just to convey a simple command. The air is thick, choking with the smell of burnt rubber, stale tobacco, and the distinct, acrid scent of unwashed bodies sweating through wool in the dead of winter.

This is the birth of the modern world. This is the Ford Motor Company's moving assembly line. And for the men standing on the line, it is hell on earth.

To understand why our current economy is collapsing, we must first understand the architecture of the cage we have been living in for the last century. We have to understand what was lost the moment Henry Ford flipped the switch.

Before this moment, a craftsman built a carriage. He touched the wood, he shaped the metal, he saw the product from start to finish. He had dignity. He had agency. His last name was likely "Smith" or "Carpenter" or "Cooper"—a name derived from his utility, yes, but also from his mastery.

But Henry Ford killed the craftsman and replaced him with the "hand".

In the Fordist system, you are no longer a maker; you are a lever of flesh. You turn Nut A onto Bolt B. You do it ten times a minute. Six hundred times an hour. Five thousand times a day. The intellectual requirement of the job was reduced to the synaptic firing speed of a reptile. If you drop a wrench, the line doesn't stop—you just get crushed, or fired, or both.

The psychological toll of this transition was immediate and devastating. The turnover rate at Highland Park in 1913 was 370%.

Read that again. 370%.

That means to keep 14,000 positions filled, Ford had to hire 52,000 men a year. The work was so soul-crushing, so monotonous, so physically destroying that men would walk off the line in the middle of a shift and never come back. They chose poverty over the machine.

This wasn't laziness. Men were not quitting because they were weak. They were quitting because they were sane. The human animal is not designed to be a metronome.

Henry Ford had a problem. He had built the most efficient engine of production in human history, but the biological components —the humans—kept breaking. He had a factory that could churn out a Model T every 93 minutes, but he didn't have a stable workforce to build them, and more importantly, he didn't have a population rich enough to buy them.

He was staring at the great contradiction of industrial capitalism: Efficiency without consumption is just bankruptcy with better robots.

Section 2: The Luddite Warning

We must pause here to correct a historical lie. The history books —written by the winners—have conditioned you to believe that resistance to this machine logic is futile and foolish. They taught you the word "Luddite."

In popular culture, a Luddite is a backward idiot who hates technology. Someone who smashes a computer because they don't know how to use email.

This is propaganda.

One hundred years before Ford, in 1811, the textile workers of Nottingham, England—followers of the mythical General Ned Ludd—did not smash weaving frames because they hated technology. They smashed them because the technology was being used to circumvent labor laws and produce inferior goods while starving the workers.

The Luddites were not anti-technology; they were **pro-human**.

They saw exactly what Ford would see a century later: that if you detach production from the well-being of the producer, you create a social bomb. The Luddites were the first to realize that the machine was not a tool, but a competitor.

The British government responded not with a Universal Basic Income or a transition plan, but with the death penalty. They deployed 12,000 troops to crush the weavers—more soldiers than Wellington took to the Iberian Peninsula to fight Napoleon.

Why does this matter in 2026? Because the Luddites were the "Canary in the Coal Mine." They were the first casualties of the war between Capital and Labor. They lost. And because they lost, the world of Henry Ford became possible.

Ford looked at the Luddite equation and realized he couldn't use soldiers to force men to work. He needed a subtler weapon. He needed a weapon that would make the men *want* to stay in the cage.

Section 3: The Golden Handcuffs

On January 5, 1914, Henry Ford changed the rules of the game. He didn't just tweak the system; he rewired the operating system of human society.

He announced the Five-Dollar Day.

At the time, the average unskilled worker made $2.34 a day. Ford more than doubled it overnight. The news hit the wire services like a bomb. It wasn't just economic news; it was a cultural earthquake.

Within 24 hours, ten thousand men were shivering outside the gates of the Highland Park plant in zero-degree weather, desperate for a job. It got so chaotic the police turned fire hoses on the crowd to beat them back. The water froze on the men's coats, turning them into ice sculptures of desperation.

They stood there, freezing, because they thought Henry Ford was a savior.

History books tell you this was the moment the American Middle Class was born. They tell you Ford was a visionary philanthropist who wanted his workers to live with dignity. They teach this in every MBA program from Harvard to Stanford as the "Virtuous Cycle".

They are lying to you.

The Five-Dollar Day was not a gift. It was a leash.

Henry Ford didn't raise wages because he had a bleeding heart. He was a ruthless industrialist who busted unions and hired spies to watch his employees' homes. He raised wages because he did the math. He realized that if he wanted to become the richest man in history, he needed to invent a new class of human being.

He didn't create the Middle Class. He created the **Consumer Class**.

Think about the architecture of the deal. By doubling the wage,

Ford accomplished two things instantly:

1. **He secured the Labor:** The turnover rate dropped from 370% to 16%. The fear of losing that $5 goldmine kept men glued to the line, tolerating the intolerance, destroying their bodies for the payout. He bought their silence.
2. **He secured the Market:** This is the darker truth. By putting money into the pockets of his workers, he gave them just enough capital to buy the very product they were building.

The money went out in the payroll envelope on Friday, and it came back to the Ford Motor Company on Monday when the worker put a down payment on a Model T.

This was **The Hamster Wheel Solution**.

Ford realized that you cannot extract wealth from a stone. If you want to sell mass-produced goods, you need a mass of people with money. But—and this is the critical part—you must never give them enough money to stop working.

You give them enough to *consume*, but never enough to *own*.

The $5 wage was calculated down to the penny. It was calibrated to be exactly enough to afford a mortgage, a car, and a suit, but not enough to buy stock in the Ford Motor Company.

This is the trap you are still living in today. Look at your own life. Look at your paycheck. It is designed to cover your subscription services, your lease, your mortgage, and your food. It flows through you like water through a pipe. You are not a vessel for wealth; you are a conduit.

You are the transmission line that moves money from the corporation (wages) back to the corporation (consumption). Ford's genius was realizing that the worker is a battery. You charge him up with a wage, and then you drain him with products. As long as the voltage stays balanced, the machine runs forever.

Section 4: The Severed Loop

For 100 years, this worked. It worked because the "Hamster Wheel Solution" relied on one fundamental axiom: **Labor is necessary**.

Ford *needed* those men. He needed their hands to turn the wrench. He needed their eyes to spot the defects. Because he needed them, he had to feed them. The wage was the maintenance cost of the biological machinery.

The "Social Contract" that the politicians weep over? The one they say is broken? It was never a contract between equals. It was a user agreement. The deal was simple: We will tolerate the drudgery of the assembly line, and in exchange, you will give us enough trinkets to feel like kings for the weekend.

But let's look at the ledger. Let's look at who actually got rich.

- The worker got a car that rusted in ten years.
- He got a house that kept him in debt for thirty years.
- He got a wage that vanished as soon as inflation ticked up.
- Henry Ford? He didn't take a wage. He took **Equity**.

He owned the factory. He owned the patents. He owned the land. When the company grew, the worker got a 5% raise; Ford got a 500% increase in net worth.

The worker was playing a game of addition ($5 + $5). The owner was playing a game of multiplication ($5 x Volume x Leverage).

Universal Truth: You cannot work your way to freedom in a system designed to monetize your fatigue.

Now, fast forward to 2026. Stand in a modern data center.

It is silent. It is cool. There are no sweating men. There are no shouting foremen. There is just the hum of fans and the blinking of blue LEDs.

The "Hamster Wheel Solution" is breaking apart, but not because the elites have had a change of heart. It is breaking because the Hamster is dead.

Henry Ford paid $5 because he needed you to build the car. OpenAI, Google, and Tesla do not need you to build the code. They do not need you to drive the truck. They are building a new loop, one where the production is automated and the costs are zero.

The Loop of 1914—Production creates Wages, Wages create Consumption, Consumption drives Production—has been severed.

The engine is revving, the pistons are firing, but the driveshaft has snapped.

They have kept the "Consumer" part of the deal (they still want you to buy), but they are deleting the "Worker" part of the deal (they don't want to pay you).

And here you are, sitting with a resume in one hand and a smartphone in the other, wondering why the old rules don't work. You are wondering why you can't find a "Good Job" with a "Fair Wage". You are angry because you think the system is malfunctioning.

It isn't malfunctioning. It is evolving.

The trap of 1914 was a brilliant lie, but it was a sustainable lie. It kept food on the table. But the lease on that lie has expired. The landlords of the future—the **Techno-Feudalists**—are looking at the Hamster Wheel, and they are looking at the AI stack, and they have realized something terrifying:

Electricity is cheaper than calories.

Ford domesticated you. The new system plans to obsolesce you.

The only way out is not to run faster on the wheel. It is to break the wheel entirely. We must stop asking for better wages. The wage is a relic of the Ford era.

We must start demanding what Henry Ford kept for himself.

Dividends.

CHAPTER 2: THE SUICIDE PACT

Section 1: The Boardroom at the End of the World

New York City. 45th Floor. Present Day.

The room is cold, soundproofed, and smells faintly of espresso and fear. This is the boardroom of a Fortune 500 company. It doesn't matter which one. It could be insurance, logistics, media, or retail. The math is the same everywhere.

Sit in the chair. Look at the faces.

These are not the mustache-twirling villains of a comic book. They are not plotting to starve the poor for sport. If you think the enemy is "evil," you are missing the point. The enemy is **Structure**.

These men and women are exhausted. They are anxious. They are trapped in a mathematical vice that is squeezing the humanity out of them drop by drop.

They are staring at a PowerPoint slide that contains two numbers.

- **Number 1: The Cost of Human Labor.** (Rising. Healthcare is up 8%. Taxes are up. Inefficiency is constant. Humans need sleep. They complain. They get sick. They organize.)
- **Number 2: The Cost of Intelligence.** (Plummeting. $0.002 per 1,000 tokens. No sleep. No unions. No soul. No friction.)

The CEO clears his throat. He adjusts his tie, which costs more than your first car. He has a **fiduciary duty** to his shareholders.

That is a fancy legal term that effectively means: *If I do not maximize profit, I will be sued and fired.*

The dilemma is brutal in its simplicity. If he fires the Customer Service department—5,000 people—and replaces them with an AI Agent that costs a fraction of a penny per interaction, the company's operating margin jumps by 15%.

The stock price will rally. The board will applaud. He will keep his job for another quarter. He will get his bonus.

But here is the catch. The CEO knows, deep down in the reptile part of his brain, that the CEO across the street is doing the exact same thing. And the CEO of the bank is doing it. And the CEO of the car manufacturer is doing it.

They are all looking at the same slide. They are all reaching for the same button.

This is the **Prisoner's Dilemma** played out with nuclear weapons.

If one company automates, they win. They get a competitive advantage. They lower their costs, undercut their rivals, and capture market share.

If *every* company automates, nobody wins. They destroy the very ecosystem that allows them to exist.

Why? Because those 5,000 people in Customer Service weren't just "costs" on a spreadsheet. They were **customers**.

The woman answering phones at the insurance company was the same woman buying diapers from the retail giant. The truck driver for the logistics firm was the same man taking out a loan from the bank. The coder for the software firm was the same guy subscribing to Netflix.

The economy is a circulatory system. Money is the blood. Wages are the heart pump that keeps the blood moving.

When you fire the human, you delete a line item on your

expense sheet, but you also delete a line item on the **global revenue sheet**. You are cutting off the oxygen supply to your own lungs.

This is **Algorithmic Cannibalism**.

Corporations are eating their own customer base to survive the quarter. They are devouring the hand that feeds them because the algorithm tells them it's the most efficient way to lower calorie expenditure.

It is a suicide pact signed in ink, validated by McKinsey consultants, and cheered by Wall Street algorithms that can't see past the next millisecond.

Section 2: The Nash Equilibrium of Hell

To understand why they can't stop, we have to look at the work of John Nash, the mathematician who won the Nobel Prize for Game Theory.

Nash proved that in a non-cooperative game, individuals will make decisions that are "rational" for themselves but disastrous for the group. This state—where everyone is doing the smart thing for themselves but the result is total chaos—is called the **Nash Equilibrium**.

We are currently trapped in a **Nash Equilibrium of Hell**.

Imagine two generals, General A and General B. They are enemies. Both have the option to build a Doomsday Device.

- If General A builds it and General B doesn't, General A wins.
- If General B builds it and General A doesn't, General B wins.
- If *both* build it, they blow up the world.
- If *neither* builds it, they live in peace.

The rational choice for the group is clearly "neither builds it." But the rational choice for the *individual* is always to build it. You cannot risk the other guy building it first. So, both build it. Both die.

Now replace "Doomsday Device" with "Artificial General Intelligence (AGI)."

The CEO in that boardroom has no choice. This is the **Tyranny of the Micro**.

If he refuses to automate—if he stands up and says, "We must protect the American worker! We will keep our human staff!"— he is not a hero. He is a martyr.

His competitors—who have no such moral qualms—will automate him into bankruptcy within six months. They will lower prices, undercut him, drive him out of business, and buy his assets for pennies on the dollar.

So he pushes the button. He signs the layoff order. He feeds the cannibal.

Universal Truth: In a system based on competition, doing the "right thing" is a death sentence.

The market selects for ruthlessness. It selects for the entity that can strip-mine value the fastest. And right now, the fastest excavator in history is AI.

The "Suicide Pact" is already underway. Every time you see a headline about a "Strategic Restructuring," or a "Pivot to AI," or a massive layoff at a profitable tech company, you are watching another signature go onto the document.

They are stripping the copper wiring out of the house to sell it for scrap, while we are still living inside.

Section 3: The Deflationary Spiral

Now, zoom out. Leave the boardroom and look at the street. This leads us to the terrifying economic paradox of the AI age: **The Deflationary Spiral**.

Traditional economics—the kind taught by professors who have never run a business—says that automation is good because it makes things cheaper. If robots make shoes, shoes become cheap. Great! Everyone can afford shoes!

This logic worked when automation simply made the worker *more efficient* (The Ford Model). The worker still had a job; he just made more shoes.

But AI doesn't make the worker efficient. It makes the worker **redundant**.

Imagine a bakery. This bakery is fully automated. Robots mix the dough, bake the bread, and stack the loaves. The cost to produce a loaf of bread drops to near zero. It's a miracle of abundance! The bread is cheaper than it has ever been in human history.

But outside the bakery window, there is a man. He used to be the baker. Now, he is unemployed.

The bread costs $0.10.

The man has $0.00.

It doesn't matter how cheap the bread is. **Zero purchasing power makes any price infinite.**

This is the nightmare scenario we are barreling toward: A world of **Super-Abundance locked behind glass**.

Warehouses overflowing with cheap, AI-generated goods, digital services, and content, sitting stagnant because the mechanism for distributing money (wages) has been dismantled.

We are building a utopia for robots and a dystopia for humans.

The "Ford Loop" (Worker -> Wages -> Consumer) was a closed circuit. The electricity flowed.

The "AI Loop" is a broken wire. (AI -> Product -> ... Silence).

Who is the customer?

This question haunts the sleep of the smart money. Who buys

the iPhone 18 when unemployment is 40%? Who subscribes to Disney+ when the gig economy collapses?

The Silicon Valley accelerationists will tell you, "Prices will fall so low that everything will be basically free!"

Don't be an idiot.

Rent will not be free. Land is scarce.

Gold will not be free.

Power will not be free.

The **Means of Survival** will still cost money. And the entities that own the AI—the Techno-Feudal lords—will not give away their product out of the goodness of their hearts. They have servers to pay for. They have energy bills to settle. They have shareholders to appease.

They will try to extract profit from a stone. And when they realize the stone is dry, the system will seize up.

Section 4: The Ghost of 1930 (Poverty Amidst Plenty)

We have seen this movie before. We saw a preview of this in the 1930s.

The Great Depression wasn't caused by a lack of production. The factories were still there. The farms were still fertile. The sun still shone. The rain still fell.

We had **too much** stuff.

In 1932, while children in Chicago were fighting over garbage scraps in alleyways, farmers in Iowa were burning corn in their furnaces because the price was too low to sell.

Dairy farmers were pouring thousands of gallons of fresh milk into the gutters of the highway, creating white rivers of waste, while mothers in New York City mixed flour with water to feed their infants.

Why? Because the "mechanism of exchange" broke.

The supply was there. The demand was there (people were hungry). But the **money**—the connective tissue—had vanished.

We are about to replay the Great Depression, but at the speed of light.

In the 1930s, the culprit was a banking collapse and a contraction of the money supply.

In the 2020s, the culprit is the Zero Marginal Cost revolution.

We are creating a world where AI can produce infinite software, infinite art, infinite legal advice, and infinite logistical planning. We are creating a "glut" of production.

But because we insist on tying "survival" to "labor," and because labor is becoming worthless, we are engineering a famine in the middle of a feast.

The "Suicide Pact" is the decision to maintain the **Wage System** in an era where wages are mathematically obsolete.

It is the refusal to open the Third Door (which we will discuss in Part III).

Section 5: The Acceleration

The CEO in that boardroom knows it. He isn't stupid. He reads the same reports I do. He knows that in five years, the value creation will be done entirely by the Neural Net.

- Who creates the value? The AI.
- Who captures the value? The Shareholder.
- Who is left out? You.

But he can't stop. The invisible hand of the market is pushing him toward the cliff edge. He is praying that someone, somewhere, figures out a solution before the bottom falls out.

He is hoping for a miracle. He is hoping for a new mandate.

Because without it, the only growth industry left will be **riot**

control.

Universal Truth: When the customer dies, the business dies. We are currently murdering the customer to save the business.

The engine is broken. The loop is severed. And the people in charge are pressing the accelerator to the floor.

They call it "Progress."

I call it "Velocity toward the brick wall."

It is time to take your foot off the gas and look under the hood. It is time to realize that "Jobs" are never coming back.

And if we want to eat the bread behind the glass, we need a new way to break the window.

CHAPTER 3: THE NEW PLANTATION (THE DIGITAL SERF)

Section 1: The Death of the Customer

Unlock your phone. Look at the grid of colorful icons. Instagram. TikTok. X. Google Maps. Amazon.

You think these are tools. You think they are services built for your convenience, designed to help you navigate the world, connect with friends, and buy toilet paper without leaving your couch. You call yourself a "User." You call yourself a "Customer."

You are wrong.

In the strict economic sense, a customer is someone who pays money in exchange for a good or service. This exchange creates a specific power dynamic: The customer has leverage. If the bread is stale, you stop buying the bread. The baker must improve or starve. The money is the vote.

When was the last time you paid for a Google search? When did you write a check to Mark Zuckerberg for the privilege of posting a photo of your dog? When did you pay the invoice for using Waze to dodge traffic?

You don't pay.

And if you are not paying, you are not the customer. You are the **inventory**.

We have not evolved past the Middle Ages; we have simply digitized them. The economic structure of the 21st Century is not Capitalism. Capitalism requires free markets, private property, and the voluntary exchange of labor for wages. That system is dying.

We have reverted to **Techno-Feudalism**.

To understand where we are, we have to look back at where we came from. We have to strip away the silicon veneer and look at the power structure underneath.

Section 2: The Feudal Retrograde

Travel back to the 12th Century. Europe is a dark, dangerous place. The Roman Empire has collapsed. There are no police. There are no laws. There is no currency. There are only wolves, bandits, and starvation.

If you were a peasant in 1100 AD, you had a problem. You needed land to grow food, and you needed walls to keep out the Vikings. But you didn't own land, and you couldn't build a castle.

So, you made a deal. You approached the local Lord—the guy with the big sword, the stone fortress, and the private army.

You struck a bargain: *I will work your land. I will grow your wheat. I will give you the majority of what I produce. In exchange, you let me live inside the walls when the barbarians come.*

This was the Feudal Contract.

Labor for Access.

You owned nothing. You had no rights. You were "tied to the land." If the Lord sold the land, he sold you with it. You were property.

Now, snap back to 2026.

The internet is the new Wild West. It is a chaotic ocean of information, cyber-crime, and infinite noise. To navigate it—to exist in the modern world—you need a platform. You need a digital identity. You need a place to stand.

But you don't own the servers. You don't own the fiber optic cables. You don't own the code. You don't own the "Cloud."

So, you make a deal. You approach the Tech Lords—Musk, Zuckerberg, Pichai, Bezos.

You strike a bargain: *I will work on your platform. I will create content. I will give you all my data. I will organize your information. In exchange, you let me have an email address so I can get a job. You let me have a profile so I can talk to my friends. You let me use the map so I don't get lost.*

This is the Techno-Feudal Contract.

Data for Access.

The "Platform" is the Land.

The "Cloud" is the Castle.

The "Terms of Service" is the Fealty Oath.

And you? You are the peasant toiling in the fields of the feed, harvesting likes and clicks for a Lord who watches from a glass tower in Silicon Valley.

Universal Truth: We are not citizens of the internet. We are sharecroppers on the estates of Big Tech.

The genius of this new system is that it is invisible.

In 1100 AD, the serf knew he was a serf. He looked at the Lord's silk robes, looked at his own mud-stained rags, and understood the hierarchy. He hated it, but he understood it.

Today, the serf thinks he is a King.

You have an iPhone (a shiny mirror). You have a "Personal Brand." You curate your profile. You think this is *your* space. You think *you* are the influencer.

Try this experiment: Stop paying your rent. What happens? You get evicted.

Now, try this: Violate the "Community Guidelines" of YouTube or Instagram. What happens? You get de-platformed.

Your entire digital existence—your photos, your memories, your business contacts, your reputation—can be deleted in one second by an algorithm you cannot see and a moderator you cannot appeal to.

You own **nothing**.

You are squatting on digital land that belongs to a corporation. You are tilling the soil of their servers, believing that the crop belongs to you.

Section 3: The Indenture Contract (The TOS)

Let's look at the contract itself. The **Terms of Service (TOS)**.

When you sign up for a new app, a box pops up. It contains 50,000 words of dense legal jargon, written in a font size designed to discourage reading. You scroll to the bottom and click "I Agree."

You treat this like a nuisance. You think it's just a formality, like a "Do Not Walk on the Grass" sign.

It is not a formality. It is an **Indenture Contract**.

If you actually read the TOS of the major platforms, you would be horrified. You are not just agreeing to follow the rules. You are signing away your civil rights.

- You are granting these companies a perpetual, irrevocable, worldwide, royalty-free license to use, reproduce, modify, and distribute your content.
- You are signing away the rights to your face (biometrics).
- You are signing away the rights to your location (GPS).
- You are signing away the rights to your voice (microphone).

- You are signing away the rights to your private thoughts (search history).

In the Middle Ages, the Lord was greedy. He demanded 30% or 40% of your grain. He left you enough to survive, because if you died, he had no one to work the field.

In the Digital Age, the Lord takes **100% of your data**.

Every single click. Every hover. Every message sent and deleted. It is all captured. It is all harvested.

And what do they do with it?

They don't just "sell it to advertisers." That is the old model. That is the 2010 model.

In the 2020s, they use it to **train the replacement**.

They take the photos you uploaded of your art and feed them to Midjourney so it can learn to paint like you. They take the code you pushed to GitHub and feed it to Copilot so it can learn to code like you. They take the essays you wrote on Reddit and feed them to ChatGPT so it can learn to think like you.

You are digging your own grave, one click at a time. And you signed a contract giving them the shovel for free.

Section 4: The Gamification of the Mines

This brings us to the nature of the "Work."

Most people don't think they work for Facebook. They think they *use* Facebook to goof off. They think they *use* Google to find a recipe.

This is the greatest psychological trick ever pulled. It is the **Gamification of Labor**.

Imagine if Henry Ford had tricked his workers into thinking that assembling the Model T was a "fun game" called *Bolt Tycoon*.

Imagine if he painted the wrenches bright colors. Imagine if, every time you tightened a bolt, a little digital confetti explosion went off and a bell chimed. Imagine if he created a "Leaderboard" showing who tightened the most bolts that hour.

And imagine if, instead of paying them $5 a day, he paid them in "Ford Points" that they could only use to decorate their station on the assembly line.

He would have had people lining up to work for free. He would have had them working 16 hours a day, addicted to the "ding" of the bolt tightening.

That is exactly what social media is.
- **"Likes"** are not social validation. They are a non-monetary currency used to incentivize free labor.
- **The "Feed"** is not a news source. It is a conveyor belt of tasks. "Judge this." "React to this." "Share this."
- **CAPTCHA** ("Click all the traffic lights") is not a security measure. It is you working for free to train a self-driving car's vision system.

Every time you verify that you are "not a robot," you are training a robot to be you.

You are working. You are creating value. You are the source of the raw material that powers the multi-trillion-dollar AI economy.

The algorithm knows exactly when to give you a Like to keep you scrolling. It knows exactly when to show you a rage-inducing post to keep you commenting. It is managing you.

It is the most efficient foreman in history, extracting maximum productivity from your brain stem. It doesn't need to shout. It just needs to nudge your dopamine receptors.

You feel this, don't you?

You feel the drain at the end of the day. You stare at the screen until your eyes burn. You feel the anxiety when you can't check your notifications. You feel the "phantom vibration" in your pocket.

They call it "Social Media Addiction." They treat it like a medical condition. They tell you to do a "Digital Detox."

I am telling you it is not addiction. It is **exhaustion**.

Universal Truth: You think you are addicted to your phone. No. You are exhausted from working a second job that you don't get paid for.

You are tired because you have been mining data for Google for 16 hours straight. You are tired because your brain is a biological GPU that has been running at 100% capacity processing information for the network.

And who gets the check?

Look at the stock prices. Meta. Alphabet. Amazon. Microsoft.

They are the most valuable companies in the history of the human race. Did they get rich because they have the best software? No. Open-source software is often better.

They got rich because they have the biggest Plantation. They have the most serfs. They have 3 billion people working for free, every single day, feeding the machine.

Section 5: The Extraction Economy

In the old economy—the Industrial Economy—wealth was generated by extracting resources from the earth. You drilled for oil. You mined for coal. You cut down trees.

In the new economy—the Extraction Economy—wealth is generated by drilling into **you**.

You are the oil well. Your behavior, your creativity, your relationships, your fears, your desires—this is the crude oil of the 21st century.

They extract it, refine it into "Prediction Products" (advertising and AI models), and sell it to the highest bidder.

This is why the services are "free." If you are not paying for the product, you *are* the product. But it goes deeper than that. You are the *resource*.

And just like the serf in 1200 AD, you are trapped.

This is the concept of **Network Effects**, weaponized as a prison wall.

You cannot leave.

If you delete your social media, you become a ghost. You lose your network. You lose your ability to stay informed. You lose your "social credit."

In a world where connection is currency, disconnection is bankruptcy.

If you try to leave the plantation—if you delete LinkedIn, delete WhatsApp, delete Gmail—you become unemployable. You become a social pariah. You are cut off from the tribe.

So you stay inside the walls. You keep clicking "I Agree." You keep scrolling. You keep working.

But the harvest is coming.

For twenty years, we have been feeding the beast. We have given it our photos, our words, and our logic. We thought we were just sharing memories.

We were actually building our replacement.

The data we gave them for free has now been used to build the Artificial Intelligence that is coming for our paid jobs. The Instagram photos trained the AI artist that replaced the graphic designer. The GitHub code trained the AI coder that replaced the software engineer.

We dug our own grave, one tweet at a time.

The Digital Serfdom is not just about privacy. It is about **Theft**.

It is about the unpaid appropriation of the collective intelligence of humanity. The Feudal Lords of Silicon Valley have claimed ownership over the output of the human race. They have put a fence around the "Commons" of human thought and started charging us rent to access it.

This is the reality of your life. You are not a free citizen of a Republic. You are a bonded laborer in a Digital Colony.

But unlike the Medieval serf, you have the power to realize what is happening. The walls of this castle are made of code, and code can be rewritten.

The first step to freedom is to stop calling yourself a User.

Start calling yourself a **Shareholder**.

Because if we are doing the work, we deserve a cut of the harvest.

CHAPTER 4: THE GREAT ENCLOSURE

Section 1: The Fences of 1773

History rhymes. And right now, it is screaming.

To understand why the internet—the greatest library in human history—is currently being locked behind a paywall, you have to look at a sheep pasture in 18th-century England.

Before the Industrial Revolution, there was a concept called **The Commons**.

These were open pastures, forests, and rivers that belonged to no one and everyone. They were the safety net of the peasant class. If you were poor, you could graze your cow on the village green. You could fish in the stream. You could gather firewood in the forest. It was a shared resource system that allowed the poor to survive outside of the cash economy.

Then came the **Enclosure Acts**.

The wealthy landowners and the emerging industrial class looked at the Commons and saw wasted profit. They saw "inefficiency." Why let a peasant graze a skinny cow for free when you could fence off the land, graze a thousand sheep, and sell the wool to the textile mills for a fortune?

So, they bribed Parliament. They passed laws allowing them to put up fences.

They erected stone walls around the forests and the fields. They claimed the land as Private Property.

Suddenly, if you grazed your cow, you were a trespasser. If you gathered firewood, you were a thief. The peasants were pushed off the land and into the squalor of the industrial cities to work in the factories.

The Enclosure Acts didn't create new land; they simply transferred ownership of the existing land from the "People" to the "Lords." They privatized the survival mechanism of the poor to fuel the wealth mechanism of the rich.

Fast forward to 2026.

We are witnessing the **Digital Enclosure Acts**.

For thirty years, the Internet was our Commons. It was the Village Green. It was built on protocols like HTTP and SMTP that were open, decentralized, and free. You could read a blog without a subscription. You could browse Reddit without an account. You could search the web without being tracked.

That era is over. Look around you. The fences are going up.

Have you noticed how the internet feels smaller? Have you noticed the "Log in to continue" pop-ups that block your screen? Have you noticed that Reddit killed its third-party apps? Have you noticed that X (Twitter) blocked non-users from seeing posts? Have you noticed the paywalls rising like fortress walls around every newspaper, every forum, and every archive?

This isn't an accident. This isn't just "bad UI."

This is a defensive perimeter.

The Tech Giants realized something terrifying in 2023. They realized that the "Open Web"—the free exchange of text and images—was not just a library.

It was a fuel depot.

The new engine of the economy is Artificial Intelligence. And unlike the steam engine, which ran on coal, the AI engine runs on **Tokenized Human Thought**.

Every blog post you wrote, every review you left on Yelp, every line of code you pushed to GitHub, every fan-fiction story you published, every photo you uploaded—that is the coal.

For years, companies like Google and OpenAI sent "crawlers" (digital spiders) scurrying across the Open Web, sucking up everything in their path. They downloaded the entire internet. They took your life's work, your jokes, your arguments, and your art, and they poured it into the furnace of their Large Language Models.

They did this for free. They called it "Fair Use."

But now, the data holders—the owners of the platforms like Reddit, StackOverflow, and the New York Times—have woken up. They realized they were being robbed. They realized that their content was being used to train the very machines that would destroy their business models.

So, the **API Wars** began.

Reddit realized: "Wait, OpenAI is using our users' conversations to teach ChatGPT how to sound human, and then charging $20 a month for it? And we get nothing?"

So Reddit slammed the gate shut. They put a price tag on their data. They told the crawlers to pay up or get out.

Twitter realized: "Microsoft is scraping our real-time news feed to train their bots?" So Musk pulled up the drawbridge.

This is the **Great Enclosure**. The Tech Lords are fighting each other over who owns the rights to *your* output.

- Reddit says they own your posts.
- Twitter says they own your tweets.
- The New York Times says they own their articles.

But nobody is asking the most important question: **What about you?**

You are the sheep that grew the wool. They are fighting over who gets to sell the sweater.

Section 2: The Metabolism of Theft

This brings us to the nature of the theft. And make no mistake: **It is the greatest heist in human history.**

The AI companies rely on a legal defense called "Fair Use." Their argument goes like this: *Humans read books and learn. A student goes to the library, reads Hemingway, and learns to write better. Our AI is just doing the same thing. Why is it illegal for a machine to read, but legal for a human?*

It is a seductive argument. It sounds logical.

It is a lie.

Here is the difference: **Scale and Substitution.**

When a human artist looks at a painting by Van Gogh and gets inspired, that is "influence." The human filters that influence through their own life experience, their pain, their joy, and their limited biological capacity. They produce something new that *honors* the original.

When a billion-dollar algorithm ingests 100,000 paintings, deconstructs the mathematical patterns of the brushstrokes, and creates a button that says "Generate Art in the Style of Van Gogh," that is not influence.

That is **digestion**.

OpenAI, Midjourney, and Stability AI did not "learn" from human culture. They **metabolized** it.

They took the sum total of human creativity—millions of books, billions of images, trillions of lines of code—and they compressed it into a commercial product.

Imagine you spent 20 years mastering the violin. You played in the subway station every day. A recording company recorded you secretly. They then built a robot that plays exactly like you, indistinguishable from you, but it can play 24 hours a day without rest.

They put the robot next to you in the subway.

The robot plays for free (or for pennies).

You starve.

Would you call that "Fair Use"? Or would you call that "Identity Theft"?

A human reads a book to become a better human. An AI reads a book to **replace the author**.

Universal Truth: If you train an AI on my copyrighted work, and that AI replaces me, you haven't just copied me. You have metabolized me.

They have distilled the essence of humanity into a liquid asset. They have turned "Culture" into "Compute."

And the valuation is sickening. OpenAI is valued at nearly $100 Billion (as of late 2024).

Where did that value come from?

Did it come from the code? The code is just a transformer architecture. It is open research. You can download the papers.

Did it come from the servers? Servers are a commodity.

It came from the **Weights**.

And the Weights came from **The Data**.

And the Data came from **Us**.

If you removed the copyrighted books, the artwork, and the forum discussions from the training data, ChatGPT would be a blithering idiot. It would be a blank slate.

We are the ghost in the machine.

But do we have equity? No.

Do we get a dividend? No.

We get the privilege of paying $20/month to access a lobotomized version of our own collective intelligence.

Section 3: Burning the Library

The "Enclosure" means that the era of the Free Internet is dead. The future is a series of walled gardens.

If you want to contribute, you must sign the TOS. If you want to access the truth, you must pay the subscription.

And outside the walls? A wasteland of AI-generated sludge, bots talking to bots, and hallucinations.

They have burned the library to keep the servers running.

There is a viral hook circulating in the underground tech forums: **"They burned the Library of Alexandria to keep us warm. Now they are burning the Internet to teach the robots how to think."**

The Commons is gone. The village green has been paved over to build a data center.

The Lords have successfully fenced off the raw material of the digital age. They have claimed ownership of the language, the art, and the logic of the human species.

But there is a flaw in their plan. A fatal flaw. A biological flaw.

They have enclosed the land, and they have built the machines, and they have stolen the data. But they forgot one thing.

The machines need a constant stream of **new** data to survive.

Section 4: The Mad Cow of the Internet (Model Collapse)

This is our leverage. This is the one card we have left to play.

It is a phenomenon known to computer scientists as **Model Collapse**.

The Tech Giants have a dirty secret. They are terrified.

They have built these god-like models—GPT-5, Gemini, Claude—by feeding them the entire history of the human internet up to roughly 2023. They fed them "Human Data."

But now, the internet is filling up with "AI Data."

By 2026, it is estimated that 90% of the text on the internet will be synthetically generated. Blogs written by ChatGPT, comments written by bots, tweets written by agents.

The snake is starting to eat its own tail.

If you train a future AI on data generated by a past AI, the model starts to go insane. It amplifies errors. It loses nuance. It hallucinates. It drifts into madness.

It's the digital equivalent of **Mad Cow Disease**—the result of feeding a species its own processed remains.

In biology, when you feed cows to cows, you get prions. You get degenerative brain disease.

In AI, when you feed GPT-4 output into GPT-5, you get The Habsburg Jaw of intelligence. You get inbred logic.

To keep the models smart, to keep the Artificial Intelligence from collapsing into a singularity of gibberish, they need a constant stream of **Fresh, Clean, Human Experience.**

They need the "Ground Truth."

And who has the Ground Truth?

You do.

You are the only entity in the loop that actually exists in the physical world.

The AI knows the word "Coffee." It can describe the chemical composition of caffeine. It can generate a photorealistic picture of a latte.

But it has never tasted the bitterness. It has never felt the caffeine hit the bloodstream. It has never burned its tongue.

That *experience*—that sensory data—is the only thing that anchors the simulation to reality.

Without us, the AI drifts off into space.

We are the Truth Anchors of the universe.

This changes the entire negotiation.

For Part I and Part II of this book, I have told you that your Labor is worthless. I have told you that your Credentials are worthless. I have told you that your role as a Consumer is dying.

But your role as a **Source of Reality** is just beginning.

The Tech Lords act like they are gods, and we are obsolete. But they are building a cathedral of logic on a foundation of our data. If we pull the foundation, the cathedral cracks.

This is the core of **Dividendism.**

We must stop thinking of "Universal Basic Income" (UBI) as a charity handout given to useless eaters to keep them from starving. That is a slave mentality.

We must shift to **"Data Dividends."**

We are not asking for a handout. We are demanding a royalty check.

We are the shareholders of USA Inc. We are the owners of the "Training Data" that powers the multi-trillion-dollar engine.

The companies are strip-mining our reality to fuel their machines. Fine. They can have the data.

But they have to pay for the extraction rights.

The economy of the future is not:

Work -> Wage -> Buy.

The economy of the future is:

Exist -> Generate Reality -> Dividend -> Live.

We are leaving the age of the "Worker." We are entering the age of the "Human."

The deflationary black hole will swallow the cost of goods. Food, energy, and plastic crap will become nearly free. But Trust—the knowledge that you are interacting with a fellow soul—will become the most expensive commodity on earth.

The billionaire of 2030 won't be the guy with the most robots. It will be the guy who has access to a room where no electronics are allowed, talking to people who are undeniably real.

You are valuable. Not because of what you do. The robot does it better.

You are valuable because of what you are.

The machine is a perfect liar.

You are an imperfect truth.

And in a world of infinite lies, the truth is the only thing worth paying for.

SYSTEM LOG: DIAGNOSIS COMPLETE

STATUS: CRITICAL FAILURE DETECTED

ROOT CAUSE: THE DECOUPLING OF PRODUCTION AND WAGES

SUMMARY:

- **The Ford Paradox:** The 20th-century economy relied on a closed loop (Worker earns -> Worker buys). AI has severed this loop. Robots produce, but robots do not buy.
- **The Luddite Truth:** The Luddites were not anti-technology; they were pro-human. They predicted that automation without redistribution leads to starvation. We are ignoring the canary in the coal mine.
- **Techno-Feudalism:** We have regressed from Capitalism (free markets) to Feudalism (walled gardens). We are no longer citizens; we are serfs tilling the data fields for the Lords of Silicon Valley.
- **The Great Enclosure:** The Tech Giants have fenced off the "Commons" of human intelligence. They are selling our own collective knowledge back to us at a subscription fee.

ACTION REQUIRED: PROCEED TO PART II FOR ECONOMIC PHYSICS ANALYSIS.

PART II: THE THEORY - ECONOMICS OF ZERO

CHAPTER 5: THE DEFLATIONARY BLACK HOLE

Section 1: The End of Scarcity

Tear up your Economics textbook. Use it for kindling. It is worse than useless; it is a map of a world that no longer exists.

For three hundred years, every economic theory from Adam Smith to Karl Marx was built on one foundational assumption: **Scarcity**.

There was only so much land. There was only so much gold. There was only so much grain. And most importantly, there was only so much *time*.

Because things were scarce, they had value. Because it was hard to make a chair, the chair cost money. Because it took years to train a doctor, the doctor earned a high wage.

Value was derived from **Friction**. The harder it was to produce something, the more it was worth.

But what happens when the friction disappears? What happens when the cost of production collapses to zero?

Welcome to the **Deflationary Black Hole**.

We are entering a period of history where the rules of supply and demand are inverting. We are moving from an Economy of Scarcity to an **Economy of Abundance**. And while "Abundance" sounds like a paradise, under our current system, it is a mathematical apocalypse for the working man.

To understand this, you need to understand the most dangerous phrase in economics: **Zero Marginal Cost**.

Don't let the jargon scare you. It is simple. There are two costs to make anything:

1. **Fixed Cost:** The money to build the factory or write the code.
2. **Marginal Cost:** The money to make *one more* unit once the factory is running.

In the old world (Ford's world), the Marginal Cost was high. To build the *second* Model T, Ford needed more steel, more rubber, and more man-hours. He couldn't just press a button and copy the car.

But in the digital world, the Marginal Cost is **Zero**.

Think about Microsoft Word. It cost Microsoft millions of dollars to write the code for the first copy of Word (Fixed Cost). But how much does it cost them to sell the second copy? Or the billionth copy?

Nothing.

Control-C, Control-V.

It costs electricity. It costs pennies. Because it costs nothing to replicate, the price eventually falls to nothing. This is why you don't pay for email. This is why you don't pay for GPS. This is why you can listen to any song in human history on YouTube for free.

Universal Truth: In a free market, price always falls to the Marginal Cost.

If it costs zero to reproduce, the price will trend to zero.

For the last twenty years, this "Deflationary Black Hole" was confined to the screen. It ate the music industry. It ate newspapers. It ate software.

But now, the event horizon is expanding. It is crossing the barrier from **Bits** to **Atoms**.

Section 2: The Three Collapses

The Black Hole is now swallowing the three pillars of the physical economy: Intelligence, Energy, and Labor.

1. Intelligence:

Two years ago, if you wanted a 500-word essay written, you had to hire a human. You had to pay for their time, their coffee, and their rent. The Marginal Cost was, say, $50.

Today, you ask an AI. The energy cost to generate that essay is $0.0003.

The price of writing has collapsed by 99.99%. Intelligence—the ability to process information, make decisions, and create logic —is becoming a commodity.

2. Energy:

Oil is a scarcity game. You have to find it, drill it, ship it, and burn it. Once you burn it, it's gone. You have to find more. The Marginal Cost is always there.

Solar and Fusion are technology games. Once you build the panel or the reactor (Fixed Cost), the sunlight is free. The Marginal Cost of the next kilowatt is near zero.

As technology improves, we are heading toward a world of essentially free energy.

3. Labor:

This is the asteroid impact. A human worker is a high-friction, high-marginal-cost machine. You need 2,000 calories a day. You need sleep. You need healthcare. You need a salary every two weeks.

A humanoid robot is a fixed-cost machine. You pay to build it once. After that, it runs on pennies of electricity. It doesn't need a salary.

If a robot can stock a shelf for a marginal cost of $0.10 an hour, and you need $20.00 an hour to survive, who gets the job?

The "Invisible Hand" of the market will strangle you. It has to. The math demands it.

Section 3: The Great Paradox

This leads us to **The Great Paradox**.

In a sane world, "Everything becoming cheap" should be a celebration. If energy is free, and labor is free, and goods are free, we should be living in the Garden of Eden. We should be sipping nectar while robots fan us with palm fronds.

But we are not living in the Garden of Eden. We are living in a system built on **Wages**.

Here is the nightmare logic of the Deflationary Black Hole:
1. Technology drives costs to zero.
2. Companies lower prices to compete.
3. To survive lower prices, companies must cut costs (fire humans).
4. Unemployment rises.
5. Wages collapse.

We are creating a world where **Products are Cheap, but Money is Impossible.**

Imagine a store full of gourmet food. The prices are incredibly low. A steak dinner is $1. A bottle of wine is $0.50.

But you are standing outside the window, starving.

Why? Because you have $0.00.

Because the job you used to do to earn money—cooking the steak, bottling the wine—is now done by the machine that made it cheap.

This is the gravitational pull of the Black Hole. It sucks value out of the labor market and compresses it into the capital market.

If you own the Robot (Capital), you win. You get the product for free and you sell it for pure profit.

If you are the Robot (Labor), you lose. You are demonetized.

We are witnessing the death of **Price** as a signal of value.

The economists will tell you this is "transitory." They will tell you that "new jobs will be created."

They are lying. They are applying 19th-century logic to 21st-century physics.

When the tractor replaced the horse, the horses didn't get "new jobs" as farm managers. They didn't "upskill."

They became pet food.

We are the horses.

We are barreling toward a future where the cost of *survival* (food, energy, goods) drops drastically, but the ability to *earn* drops even faster.

This isn't a recession. This is a phase change. Ice turning into steam.

The Black Hole does not care about your mortgage. It does not care about your student loans. It obeys only the laws of efficiency. And the most efficient state for an economy is one where humans are removed from the production loop entirely.

Section 4: The Death of Profit (Suicide of Competition)

Walk into a business school. Listen to the professor in the tweed jacket. He will tell you that the Holy Grail of capitalism is "Perfect Competition."

He will draw a chart on the whiteboard where Supply meets Demand. He will say that in a perfect market, information is free, barriers to entry are low, and the consumer gets the best possible price. He smiles when he says this. He thinks this is Utopia.

He is an idiot.

For a business owner, Perfect Competition is not Utopia. **It is Hell.**

In a "Perfect Market," you have no pricing power. If you charge $1.00 more than your neighbor, you lose 100% of your customers. So you lower your price. Your neighbor lowers his. You keep cutting until you are both selling your product at the exact cost of production.

Profit = $0.00.

For two centuries, friction saved us. It was hard to compare prices. It was hard to switch banks. It was hard to find a new supplier. That friction allowed businesses to charge a premium. That premium was called "Margin." And Margin is what paid your salary.

AI incinerates Friction.

We are entering the era of **Hyper-Competition**.

Imagine you start a digital marketing agency. You use AI to write copy and generate ads. You charge $1,000 a month.

Five minutes later, an AI Agent scans the web, sees your business model, copies it, and offers the exact same service for $900.

Then another Agent offers it for $800.

Then $500.

Then $100.

This doesn't happen over years. It happens in milliseconds. The algorithms have no ego. They have no overhead. They don't need to pay for a kid's braces. They will bid the price down to the absolute mathematical floor: the cost of the electricity required to run the server.

This is the **Suicide of Competition**.

When everyone has access to the same Super-Intelligence, "Intelligence" ceases to be a competitive advantage. It becomes a commodity.

If everyone is a genius, no one is.

If everyone can write code, software is free.

If everyone can generate art, art is worthless.

This creates a terrifying new reality for the economy: **The Death of Profit.**

But wait—you see the stock market hitting record highs. You see the tech giants minting billions. How can profit be dead?

Here is the secret. Profit isn't dying *everywhere*. It is migrating. It is fleeing the surface and retreating to the core.

Section 5: The Great Bifurcation (Kill Zone vs. God Tier)

The economy is splitting in two.

Layer 1: The Commodity Layer (The Kill Zone)

This is where you live. This is where the apps are. This is where the services are. This is writers, coders, designers, lawyers, accountants, and small business owners.

In this layer, AI is the predator. It creates infinite supply. Infinite supply crashes prices.

If you are building an "AI Wrapper" (a business built on top of ChatGPT), you are building a sandcastle at high tide. The water is coming. You will be competed down to zero.

Layer 2: The Infrastructure Layer (The God Tier)

This is where the profit is hiding.

If the AI Agents are fighting a price war to sell services for pennies, who wins?

- The entity that sells them the **Compute**.
- The entity that sells them the **Energy**.
- The entity that owns the **Data Center**.

Think about the math. Agent A and Agent B are fighting to the death. They are slashing prices. But to fight, they both need to burn electricity. They both need to rent GPU cycles from Nvidia or Amazon or Microsoft.

The combatants go broke. The arms dealer becomes a Trillionaire.

This is the **"Shovel Strategy"** of 1849, updated for the silicon age.

During the California Gold Rush, thousands of men flocked to the hills. They dug in the dirt. They slept in mud. Most of them died broke, holding a pan full of dust.

Who got rich? Samuel Brannan.

He didn't dig for gold. He bought every shovel, pickaxe, and pan in San Francisco, and then he sold them to the fools heading for the hills.

Today, AI is the Gold. Everyone is trying to find the "Killer Use Case." Everyone is trying to build the next big App.

You are the Gold Digger.

Nvidia is the Shovel Seller.

The "Shovel" of 2026 is not made of wood and steel. It is made of three things:

1. **Chips (Compute):** The silicon brains that process the math.
2. **Watts (Energy):** The electricity that powers the brains.
3. **Cooling (Water/Land):** The infrastructure to keep the

brains from melting.

These are physical constraints. You cannot "print" more electricity. You cannot "download" a new power plant. Because they are scarce, they hold value.

Everything else—code, text, media, services—is digital. It can be copied. Therefore, its value trends to zero.

This is why you see Microsoft signing deals to restart nuclear power plants. This is why Amazon is buying land next to hydroelectric dams.

They know the truth.

They know that the "Service Economy" is a race to the bottom. They are securing the Base Layer. They are building the toll roads that the robots must drive on.

Now, look at your own position.

Are you selling a Service? Or are you owning Infrastructure?

If you are a freelancer, a consultant, or an employee, you are selling a Service. You are in the Kill Zone. You are competing against an algorithm that gets faster and cheaper every day.

You are fighting a gravity war against a Black Hole.

The "Death of Profit" means that for 99% of businesses, margins will evaporate. We will see a massive consolidation. Millions of small businesses will vanish, replaced by a few mega-platforms that run on automated margins so thin that no human could survive on them.

Viral Hook: Capitalism is an eating contest where the winner eats the losers, and then eats itself. AI just sped up the metabolism.

We are moving toward a world where the only way to make money is to charge rent to the machines. If you don't own the chips, the land, or the power, you are just fuel for the fire.

The system is cannibalizing the Creator Class to feed the Infrastructure Class.

The writer starves so the server farm can feast.

This is the end of the "Middle Man." And unfortunately, the Middle Class was essentially a class of Middle Men—people who sat between a problem and a solution.

AI connects the problem and the solution directly.

The Middle Man is deleted.

So, the Profit is gone from labor. It is gone from services. It is pooling in the Infrastructure.

But there is one final layer to the Zero Theory.

If the value of "Artificial Intelligence" drops to zero because it is abundant, what happens to the value of "Biological Intelligence"?

What happens when the Fake becomes free?

The Real becomes priceless.

CHAPTER 6: THE SINGULARITY OF VALUE

Section 1: The Tsunami of Sludge

Fast forward three years. Open your laptop. What do you see?

You don't see the internet. You see a landfill.

You see a news article written by an AI, summarizing a video generated by an AI, commenting on a tweet written by a bot farm in St. Petersburg. You open your email. It is not just spam; it is hyper-personalized phishing written by an intelligence that scraped your mother's maiden name and your dog's birthday from a database breach five years ago.

You get a FaceTime call. It looks like your daughter. It sounds like your daughter. It uses her inflection, her slang, her laugh. It is asking for money because of an "emergency." It is a deepfake generated in real-time by a scraper that stole her voice from a TikTok video she posted in 2023.

We are approaching the event horizon of the **Deflationary Black Hole**.

This is the moment where the cost of generating *bullshit* hits absolute zero.

When the cost of creating information drops to nothing, the supply goes to infinity. We are about to be drowned in a **Tsunami of Sludge**.

For the last decade, the "Dead Internet Theory" was a fringe conspiracy theory on 4chan. It claimed that the majority of internet traffic was bots. In 2026, it is no longer a theory; it is the weather report.

The "Public Square" is gone. It has been flooded with sewage.

In this world of infinite, perfect, synthetic noise, the fundamental laws of value are about to flip.

Section 2: The Inversion of Luxury

Welcome to **The Singularity of Value**.

For the last twenty years, we have valued the "Digital." We chased the pixel-perfect image. We obsessed over 4K resolution, auto-tuned vocals, and CGI superheroes. We chased the artificial because it was clean, efficient, and new.

But in a world where "Perfect" is free and abundant, "Perfect" becomes worthless.

If an AI can generate a flawless symphony in three seconds for the cost of a blink, nobody cares about the symphony. The awe evaporates. If an AI can write a perfect sonnet, the sonnet loses its soul.

So, what becomes the premium asset? What becomes the new Gold?

The Flaw.

The Struggle.

The Biological Reality.

We are witnessing a massive inversion of the hierarchy of value.

- **Artificial = Infinite = Cheap.**
- **Biological = Finite = Expensive.**

Think about diamonds. If diamonds rained from the sky like hail, clogging your gutters and cracking your windshield, we would sweep them into the sewer. We would value a rough piece of wood more than a flawless diamond, simply because the wood was rare.

We are the wood.

This brings us to the concept of **Proof of Humanity**.

In the near future, the ultimate luxury good will not be a Louis Vuitton bag or a Ferrari. It will be a piece of text, an image, or a conversation that you can cryptographically prove was created by a human being who bled, cried, and sweated to make it.

Why? Because the machine can fake the result, but it cannot fake the **cost**.

Section 3: Skin in the Game (The Consequence)

The reason we value human output is not just "quality." It is **Consequence**.

An AI can simulate pain. It can write: "I am heartbroken." It can generate a tear on a digital cheek.

But it has no nervous system. It has no mortality. It cannot die.

Therefore, it has no **Skin in the Game**.

When a machine makes a decision, it is calculating probabilities. When a human makes a decision, it is gambling with its existence.

That "Stake"—that risk of loss, that capacity for suffering—is the chemical X that gives art, logic, and leadership its value.

Consider the Pilot Analogy.

By 2030, autopilot systems will be statistically safer than human pilots by a factor of ten. They will react faster. They will never get tired. They will never get drunk.

But when you step onto a plane, you will still want a human in the cockpit.

Why?

You trust a human pilot not because they calculate faster than the autopilot (they don't), but because if the plane goes down, **they die too**.

Their incentives are perfectly aligned with yours. They are anchored to the same physical fate. The AI pilot has no fate. If it crashes, it is just a deleted instance on a server. It feels nothing.

This applies to everything.

You trust a human artist because their work is a distillation of a life that has an expiration date. You trust a human leader because they have to live in the city they govern.

The AI is a tourist in reality. **We are the natives.**

Section 4: The Truth Premium

This leads to the final economic shift of the Zero Era: The monetization of **Trust**.

In the 20th century, you paid for Goods. You bought the toaster. You bought the car.

In the early 21st century, you paid for Services. You paid for the ride (Uber). You paid for the music (Spotify).

In the Dividend Era, you will pay for **Reality**.

The "Deflationary Black Hole" will swallow the cost of goods. Food, energy, and plastic crap will become nearly free. But Trust —the knowledge that you are interacting with a fellow soul— will become the most expensive commodity on earth.

The billionaire of 2030 won't be the guy with the most robots. It will be the guy who has access to a room where no electronics are allowed, talking to people who are undeniably real.

The "High-Touch" economy will explode.

- **Tier 1 (Free):** AI Therapy. AI News. AI Art. AI Companionship. (For the poor).
- **Tier 2 (Expensive):** Human Therapy. Human Journalism.

Human Art. Real Friends. (For the rich).

We are seeing the bifurcation of society into those who consume the **Simulation** and those who can afford the **Real**.

This is why "Data Dividends" are not just about money. They are about recognizing that **Humanity itself is the asset.**

You are valuable. Not because of what you do. The robot does it better.

You are valuable because of what you are.

The machine is a perfect liar.

You are an imperfect truth.

And in a world of infinite lies, the truth is the only thing worth paying for.

This concludes **Part II: The Theory**. We have diagnosed the disease (The Broken Engine) and defined the physics of the new world (The Economics of Zero).

The cage is visible. The trap is set. The scarcity of goods is ending, but the scarcity of truth is beginning.

Now, we must build the Solution. How do we force them to pay? How do we rewrite the laws of capitalism to recognize "Data" as "Labor"?

It's time to stop analyzing the cage and start bending the bars.

SYSTEM LOG: PHYSICS UPDATE

STATUS: DEFLATIONARY SPIRAL IMMINENT

SUBJECT: THE ECONOMICS OF ZERO

SUMMARY:

- **Zero Marginal Cost:** In a digital world, the cost of copying excellence is zero. Therefore, the price of everything trends toward zero. Profit evaporates for everyone except the platform owners.
- **The Great Paradox:** We are entering a world of "Poverty amidst Plenty." We will have cheap goods but no money to buy them.
- **The Kill Zone:** Middle-class jobs (services, administration, basic coding) are in the "Kill Zone." They will be automated first.
- **The Singularity of Value:** As the "Artificial" becomes infinite and free, the "Biological" (Trust, Empathy, Imperfection) becomes scarce and expensive.

ACTION REQUIRED: PROCEED TO PART III FOR THE PATCH INSTALLATION.

PART III: THE SOLUTION - DIVIDENDISM

CHAPTER 7: THE THREE DOORS

Section 1: The Nash Bargain

Pull up a chair. Sit down.

Imagine a long mahogany table in a room with no windows. The air is recycled, cold, and smells of ozone.

On one side of the table sit the Trillionaires. The Tech Lords. The owners of the algorithms, the servers, and the energy. These are the men who have successfully captured the future. They hold the keys to the "God-Mind"—the AI that can out-think, out-produce, and out-maneuver any human being.

On the other side of the table sits... You. **The American People.**

You look tired. You are holding a resume that no one reads and a bank statement that bleeds red. But under the table, your hand is resting on a detonator.

The doors are locked. There is no leaving until a deal is signed.

For the last 100 years, we played a specific game called "Labor for Wages". It was a cooperative game. You worked, they paid, you bought, they profited. The loop was closed.

We've established in Part I and Part II that the game is over. The board has been flipped. The loop is severed.

Now, we are in a **Mexican Standoff**.

- **They have the Capital.** They have the AI, the factories, the drones, and the surveillance state. They can starve you.

- **You have the Chaos.** You have the numbers. You have the pitchforks. You have the ability to burn the data centers to the ground. You can bankrupt (or hang) them.

This is a classic Game Theory problem. The current path leads to mutual destruction. The Elites know this. That is why they are building bunkers in New Zealand. That is why they are buying private islands. They can smell the smoke.

We have arrived at a fork in the timeline of human civilization. There are only three ways this story ends. There are only three doors leading out of this room.

We need to look at them coldly. We need to analyze them not with morality, but with mathematics.

Section 2: Door #1 - Elysium (Hyper-Feudalism)

This is the default path. This is what happens if we do nothing.

In this scenario, the Tech Lords look at the spreadsheet and make a cold calculation. They decide that they don't need us anymore. They have the robots to grow the food, build the houses, and guard the gates.

They look at the "surplus population"—the billions of humans who can no longer compete with AI—and they decide to cut the rope.

They retreat into their gated communities, their private cities, and their orbital space stations. They build high walls topped with automated turrets. They use AI surveillance to crush any dissent before it starts.

The economy bifurcates completely.

- **Inside the wall:** Infinite luxury, life extension technology, and post-scarcity abundance.
- **Outside the wall:** Favelas, starvation, and a *Mad Max* fight for scraps.

It sounds appealing to a sociopath. It appeals to the "Sovereign

Individual" fantasy of Silicon Valley libertarians. But there is a fatal flaw in Door #1. Two flaws, actually.

1. The Economic Suicide:

If you fire everyone, who buys the product?

Amazon is a trillion-dollar company because millions of ordinary people buy toilet paper and chargers. If 90% of the population has zero income, Amazon's revenue drops to zero. The stock market evaporates.

The wealth of the billionaires—which is mostly paper wealth based on stock valuation—vanishes.

You cannot be the King of Commerce in a graveyard.

2. The Security Paradox:

You can build a bunker. You can hire guards. You can buy robot dogs.

But you can never relax.

History teaches us one lesson: Walls always fall.

Eventually, the guards turn on you. Eventually, the mob finds a way in. Eventually, the resentment boils over into a violence so total that no amount of technology can stop it.

And specifically in the United States, Door #1 faces a logistical nightmare that does not exist in China, Russia, or Europe.

It is the **American Variable**.

The United States has 330 million people and 400 million privately owned firearms. It is the most heavily armed civilian population in the history of the species. You cannot enforce a Techno-Feudalist serfdom on a population that is better armed than most standing armies.

Furthermore, the US economy is unique. It is driven 70% by domestic consumption. In an export economy (like Saudi Arabia regarding oil), the elites don't need the population; they just need the resource. In the US, the population *is* the resource.

If you lock 300 million Americans out of the economy, you don't just get a revolution; you get an immediate collapse of the S&P 500. You cannot shoot your own customers and expect your Amazon stock to keep rising.

Therefore, in America, the Nash Equilibrium is not a choice. It is a hostage situation where both sides hold a gun to the other's head. The Dividend is the only way to lower the hammers.

Do you really want to live your life looking over your shoulder? Do you want to live in a gilded cage, terrified of the people outside?

Door #1 is not a victory for the Elites. It is a high-stress prison sentence followed by a violent death.

Section 3: Door #2 - The Matrix (The Welfare Zoo)

This is the path the politicians prefer. It is the path of least resistance.

They call it **Universal Basic Income (UBI)**.

Do not confuse what I am proposing (The Dividend) with what they are proposing (The Allowance). The difference is everything.

In Door #2, the Elites realize they can't let you starve (because of the risk of Door #1), but they don't want to give you power.

So, they print money.

They give every citizen a monthly stipend. Let's say $2,000 a month. Just enough to rent a pod, eat bug-protein paste, and pay for your subscription to the Metaverse.

They sedate you. They legalize all drugs. They make video games hyper-realistic. They create a digital fantasy world to keep you distracted while they own the physical real world.

You become a pet. You are a panda in a zoo, fed by the zookeeper, kept alive because it looks bad to let you die, but stripped of all agency, purpose, and dignity.

But here is the mathematical flaw of Door #2: **Inflation**.

If the government prints money to give to you, but you produce nothing in return, the value of the currency collapses. The cost of living will rise exactly as fast as the UBI check.

The $2,000 check will arrive, but a loaf of bread will cost $50.

You will be trapped in a permanent state of subsistence. You will own nothing. You will be "Happy," but you will not be free.

And eventually, the human spirit revolts against the zoo. A society with no purpose, no struggle, and no ownership is a society that rots from the inside out.

Door #2 is a slow, suffocating death.

Section 4: Door #3 - The Dividend Mandate (The Nash Equilibrium)

This is the only door that stays on the hinges.

In Game Theory, a "Nash Equilibrium" is a state where no player can improve their position by changing their strategy. It is the "Win-Win" that stops the war.

Here is the deal we put on the negotiation table: **The Grand Bargain**.

To the Elites, we say:

"You keep the companies. You keep the factories. You keep the algorithms. We will not seize the means of production (Socialism). We will not break up the monopolies (Anti-Trust). We will let you run the machine at maximum efficiency. We will let you automate every job. We will deregulate the speed of AI."

In exchange, you give us:

"A 20% equity stake in USA Inc. Not a tax. A Dividend."

"We acknowledge that our Data and our Infrastructure are the 'Silent Partners' in your business."

"Therefore, every quarter, a percentage of the gross profits of the Automated Economy is distributed directly to every citizen-shareholder."

Why is this the only rational choice?

For You (The Citizen):

You get income that is tied to production, not printing.

As the AI gets smarter and the economy grows, your check gets bigger. You are not a charity case; you are an owner.

You get the "Freedom Floor" under your feet.

For Them (The Elites):

This is the hardest sell, but the most important one. Why would Jeff Bezos or Elon Musk agree to give away 20% of the upside?

1. It Solves the Demand Problem.
    By putting money into the hands of the people, they create billions of customers who can afford to buy the services of the AI. The velocity of money accelerates. The economy explodes upward. 80% of a watermelon is worth more than 100% of a grape.
2. It Solves the Guillotine Problem.
    By making every citizen a shareholder, you turn every citizen into a capitalist.
    If I own shares in the AI, I don't want to smash the robot. I want the robot to work harder, because it pays me.
    You eliminate the class war instantly. You align the incentives of the Billionaire and the Single Mom. Both want the economy to succeed.

This is **Guillotine Insurance**.

The cost of the Dividend is a fraction of the cost of the security state required to maintain Door #1. It is a fraction of the cost of the economic collapse of Door #2.

We need to look the Elites in the eye and say:

"You have won the game of Capitalism. Congratulations. You possess all the chips. But the game cannot continue if the other players have no chips left to bet. The game ends. The table gets flipped."

"If you want to keep playing—if you want to stay rich, stay powerful, and stay alive—you have to 'ante up' for the next round."

"You have to buy us back in."

Universal Truth: You can pay the Dividend, or you can pay the Guards. The Dividend is cheaper.

This isn't Communism. Communism is state ownership.

This isn't Feudalism. Feudalism is lord ownership.

This is Dividendism. It is Universal Private Ownership.

It is the recognition that in an age of automated abundance, the "License to Operate" comes with a fee. And that fee is payable to the people who provided the raw materials (Data) and the environment (Society) that made the abundance possible.

The Elites are smart. They are calculators. They will eventually realize that Door #3 is the only math that balances the equation.

But they won't open it out of kindness. They will only open it if we weld Door #1 and Door #2 shut. They need to know that we **refuse** to starve (Door 1) and we **refuse** to be pets (Door 2).

We are standing at the threshold. The "Three Doors" are the only options left. The old world is gone.

The only question remaining is: How do we structure this deal? How do we mechanically turn a nation of workers into a nation of shareholders without crashing the system?

We need a mechanism. We need a "Sovereign Wealth Fund." We need to treat the United States not like a country, but like a Corporation.

In the next section, we draw the blueprints for **USA Inc.**

CHAPTER 8: THE AMERICAN SOVEREIGN FUND

Section 1: The Alaska Prototype

Stop thinking like a taxpayer. Start thinking like a shareholder.

For the last century, our relationship with the government and the economy has been based on **extraction**. You work; the government takes a cut (Income Tax). You buy; the government takes a cut (Sales Tax). A corporation makes a profit; the government tries—and usually fails—to take a cut (Corporate Tax).

It is an adversarial system. It is a game of Hide-and-Seek. The corporations hire armies of accountants to hide the money in Ireland, Luxembourg, or the Cayman Islands. The government hires armies of bureaucrats to find it.

It is inefficient, it is slow, and in the age of AI, it is obsolete. You cannot tax a fluid, digital, hyper-speed economy with a tax code written in 1954.

We need a new operating system. We need to convert the United States from a debt-ridden bureaucracy into a profit-sharing corporation.

We don't need to invent this idea from scratch. We already have a working prototype. Look at **Alaska**.

In 1976, the state of Alaska did something revolutionary. They discovered massive oil reserves on the North Slope, specifically in Prudhoe Bay. The politicians had a choice. They could have done what every other government does: tax the oil companies, take the cash, and spend it on "projects" that line the pockets of their friends—building bridges to nowhere and bloating the bureaucracy.

But Governor Jay Hammond had a different idea. He argued that the oil didn't belong to the oil companies, nor did it belong to the politicians.

The logic was simple: **The oil belongs to the people of Alaska**. The company is just the contractor extracting it.

So, they passed a constitutional amendment to create the **Alaska Permanent Fund**. They mandated that a percentage of the oil royalties be deposited into a sovereign wealth fund. That fund invests the money in stocks, bonds, and real estate. And every year, every resident of Alaska—man, woman, and child—gets a check.

They don't call it welfare. They don't call it a handout. They call it a **Dividend**.

For nearly 50 years, Alaskans have received this check. It ranges from $1,000 to $3,000 per person per year. It has significantly reduced poverty. It has kept the economy fluid. And most importantly, it is politically untouchable. Any politician who suggests touching the Dividend is immediately voted out of office.

This is the proof of concept. It proves that you can have a system where the "Commons" (natural resources) pays a dividend to the citizenry without destroying capitalism.

Now, scale that up.

We don't have enough oil to fund the entire United States. The shale fields of Texas and the Dakotas are great, but they aren't enough to replace the wage system. But we have something better.

We have the **New Oil**.

We have the Data. We have the Compute. We have the most valuable artificial intelligence infrastructure on the planet.

Google, Microsoft, Amazon, Meta, Nvidia, Tesla—these are the "Oil Wells" of the 21st century. They are extracting value from the American digital landscape. They are drilling into the bedrock of our culture and pumping out liquid intelligence.

So, how do we capture that value without destroying it?

Section 2: The Grand Swap

The current strategy of the Left is "Tax the Rich." They want to increase the corporate tax rate to 30%, 40%, or 50%.

This is a fool's errand.

If you raise taxes on a digital corporation, they will simply move their IP to a server in Singapore. They will hire more lawyers. They will lobby for more loopholes. You are trying to catch water with a net.

We propose a different deal. We propose **The Grand Swap**.

This is the deal we offer the Trillionaires. It is the deal that ends the class war.

The Carrot (What Corporations Get):
- **0% Corporate Tax.** That's right. Zero. No more offshore accounts. No more "Double Irish with a Dutch Sandwich" tax avoidance schemes. You keep 100% of your cash flow.
- **0% Capital Gains Tax.**
- **Radical Deregulation.** We cut the red tape. We let you build the nuclear plants, the data centers, and the drone networks at maximum speed.

The Stick (What Corporations Pay):
- **The Equity Transfer.** In exchange for the zero-tax

environment, every publicly traded company in the United States must transfer **20% of its equity (shares)** into the **American Sovereign AI Fund**.

We stop trying to tax the *flow* of money (which is easy to hide). We take ownership of the *source* of money (the stock).

This creates the largest Sovereign Wealth Fund in human history.

The Fund would hold 20% of Apple, 20% of Google, 20% of the entire S&P 500. It would effectively become the largest shareholder in the world.

This is not "seizing the means of production" in the Bolshevik sense. We are not sending commissars to run the factory. We are not putting bureaucrats on the board of directors.

We want Jeff Bezos running Amazon. He's good at it. We want the capitalists to do what they do best: ruthless optimization.

We just want 20% of the upside because they are building their empire on our land, using our data, protected by our military.

The Unicorn Clause (The Private Loophole)

But the skeptics will ask: "What about the private giants? What about OpenAI, SpaceX, or Stripe? They aren't on the stock market. Does that mean they escape the mandate?" Absolutely not. If a company remains private but reaches a valuation of over $1 Billion (Unicorn Status), the logic remains the same. We do not force them to go public (IPO). We do not force them to dance for Wall Street. Instead, they issue **Preferred Non-Voting Stock** (or Warrants) directly to the Sovereign Fund. The Fund sits on their Cap Table (Capitalization Table) just like a Venture Capitalist.

- If OpenAI stays private forever and pays dividends to Sam Altman? The Sovereign Fund gets 20% of those dividends.
- If OpenAI eventually sells or IPOs? The Sovereign Fund

cashes out its 20% alongside the founders.

- If they try to stay private just to hoard cash? The **Compute Tax** (Chapter 9) catches them on the operational side.

Remember: Public or Private is just a legal status. **Scale** is the reality. If you use American infrastructure to build a billion-dollar asset, the American people get equity. Period.

Think of it like a landlord. The landlord doesn't run the restaurant that rents his building. He doesn't tell the chef how to cook the steak. He just collects the rent.

The American People are the Landlords of the Digital Age.

Section 3: The Compound Interest Machine

Why is Equity better than Tax?

Because **Taxes are finite, but Equity is infinite.**

When the government collects a tax dollar, it spends it immediately. It goes to pay the interest on the debt, or it goes to a contractor, or it disappears into the black hole of the Pentagon. It is a "use-it-and-lose-it" resource.

Equity is a **Compound Interest Machine.**

By owning the assets, the American people benefit from the compound growth of the AI economy.

If AI creates a 10x explosion in GDP over the next decade (as predicted by Altman, Kurzweil, and every major economist), the value of the Sovereign Fund grows 10x.

Your monthly check grows 10x.

We are not fighting for a slice of a shrinking pie. We are baking a pie that expands to infinity.

Consider the "Stock Buyback" loophole. Currently, companies use their profits to buy back their own shares, which drives up the stock price. This benefits the CEO and the hedge funds, but does nothing for the worker or the tax base.

Under the Grand Swap, if a company does a stock buyback, the value of the Fund's shares goes up. The American People benefit directly from the financial engineering of Wall Street.

If they issue a dividend, the Fund gets paid.

The government stops being a "Parasite" that drags on the economy and starts being a "Silent Partner" that benefits from growth.

This aligns the incentives of the entire nation.

Right now, when Amazon automates a warehouse and fires 10,000 workers, the workers riot and the shareholders cheer. It is a Zero-Sum Game.

Under the Grand Swap, the Fund owns 20% of Amazon. When Amazon cuts costs and profits soar, the Fund's value goes up. The Dividend check goes up.

Suddenly, the worker isn't just a worker. He is an owner. He benefits from the efficiency of the robot.

We stop fighting the future. We ride the future.

Section 4: The Mechanism of Delivery

How do we distribute this? Do we trust the IRS to mail checks? Absolutely not.

We need a friction-less delivery system.

1. The Citizen Share:

Your Social Security Number (SSN) is no longer just a tracking number for your debts. It becomes your Stock Certificate. Every citizen gets one share of USA Inc. This share is non-transferable (you can't sell it to a hedge fund to pay for a bender). It is yours by birthright.

2. The Digital Wallet:

The Federal Reserve creates a direct digital wallet for every citizen. No middleman banks. No fees.

Every quarter, the dividends from the S&P 500 flow into the Sovereign Fund. The Fund takes a small management fee (0.1%) and distributes the rest instantly to 330 million wallets.

3. The Freedom Floor:

This creates a baseline income. In the early years, it might be small—perhaps $500 a month. But as the "Deflationary Black Hole" drives the cost of goods down, and the "Compound Interest Machine" drives the value of the Fund up, the purchasing power of that dividend expands.

This solves the "Deflationary Black Hole" paradox we discussed in Chapter 5.

- The robots drive the cost of goods down to zero.
- The Fund drives the income of the people up.

Low Prices + High Passive Income = The Golden Age.

Section 5: The Inevitable Critique

The critics will scream: "This is socialism! This is nationalization!"

They are wrong.

Socialism is when the government runs the company. It's the DMV running Google. That is a nightmare.

Dividendism is when the government owns a minority stake but lets the capitalists run the company.

It is **Universal Private Ownership**.

It is the realization that in a digitized world, data is a natural resource, just like oil. And just like the citizens of Alaska deserve a cut of the oil, the citizens of America deserve a cut of the data.

The Elites will fight this at first. They will whine about dilution.

But remind them of Door #1 and Door #2.

Remind them that 80% of a thriving, stable, high-growth

economy is worth infinitely more than 100% of a burning ruin.

We are offering them the deal of a lifetime. We are offering them the "License to Operate" in the 21st century without looking over their shoulders for the pitchforks.

We are turning the "Class War" into the "Shareholder Meeting".

Universal Truth: Don't tax the golden goose. Own 20% of the eggs.

The mechanism is simple. The precedent exists. The math works.

But there is one final piece of the puzzle. We have the "Equity" (The Fund). But what about the "Cash Flow"? How do we fund the government's daily operations—the military, the courts, the roads—without income tax?

How do we tax the robots directly?

We need to stop taxing Labor (which is dying) and start taxing the only thing that measures real work in the AI age.

Compute.

SYSTEM LOG: PATCH INSTALLED

STATUS: SOLUTION ARCHITECTURE DEPLOYED

SUBJECT: DIVIDENDISM (CAPITALISM 2.0)

SUMMARY:

1. **The Three Doors:** We face three futures: The Bunker (Elysium), The Zoo (UBI), or The Shareholder State (Dividendism). Only the third option preserves human dignity.
2. **The Grand Swap:** Corporations trade 20% Equity for 0% Tax. This aligns the incentives of the AI (Growth) with the incentives of the People (Survival).
3. **The Sovereign Fund:** We treat the nation like a corporation. Every citizen is a shareholder. The dividend is not welfare; it is a royalty check for our data.
4. **The Compute Tax:** We stop taxing human sweat (Income Tax) and start taxing machine friction (Watts & FLOPS). The machine pays for the civilization it disrupts.

ACTION REQUIRED: PROCEED TO PART IV FOR INDIVIDUAL SURVIVAL PROTOCOLS.

CHAPTER 9: THE COMPUTE TAX

Section 1: The Rotting Code of 1913

It is April 15th. Look at the form on your desk. The 1040. The symbol of American civic duty.

It asks you for your wages. It asks for your "Earned Income." It demands a percentage of the sweat you dropped, the hours you lost, and the stress you endured to keep a roof over your head.

You hate this form. You should. It is a relic. It is a ghost from a dead world.

The United States Income Tax was established in 1913 with the ratification of the 16th Amendment. To understand why it is failing today, you have to look at the world of 1913.

It was a world of muscle. If you wanted to build a bridge, you needed 1,000 men with rivets and hammers. If you wanted to harvest a field, you needed 50 men with scythes. If you wanted to run a bank, you needed a floor full of clerks scribbling in ledgers.

In 1913, Human Labor was the primary engine of value creation. The human being was the atom of the economy. Therefore, if the government wanted to fund itself, it made sense to tax the human.

But we are not in 1913. We are standing at the edge of the **Zero Point**.

In the age of the Zero Theory, Human Labor is no longer the

engine. It is the backup generator.

We are entering a world where a single server rack in a chilled room in Northern Virginia does the work of 10,000 accountants, 50,000 translators, and 100,000 illustrators. That server rack works 24 hours a day. It never gets sick. It never needs maternity leave. It never strikes.

And do you know how much income tax that server rack pays?

Zero.

The robot steals your job, keeps your salary, generates massive profits for the corporation, and contributes absolutely nothing to the public purse.

This is the recipe for the collapse of the State.

If the government relies on taxing payrolls, and payrolls go to zero, the government goes bankrupt. The mathematics are undeniable.

- The roads crumble.
- The fire department closes.
- The military dissolves.
- The social safety net snaps.

The politicians are currently trying to fix this by hiring 87,000 new IRS agents to audit *you* harder. They are trying to squeeze more blood from a stone that is already dry.

They are looking in the wrong place. We need to stop taxing the human. We need to implement **The Compute Tax**.

Section 2: The Shell Game (Why Money is a Liar)

Why not just tax the "Profits" of the AI companies? Why not just raise the Corporate Tax rate?

Because financial accounting is fiction. It is a shell game played by lawyers and accountants who are smarter than the politicians.

In the globalized digital economy, "Profit" is a fluid concept. It is like trying to nail Jell-O to a wall.

Let's say Google makes $10 Billion in profit in the United States. They don't want to pay 21% tax on that. So, their accountants use a strategy called "Base Erosion and Profit Shifting" (BEPS).

They say: "Actually, the 'Intellectual Property' for our AI is owned by a subsidiary in Bermuda. So, the US branch has to pay a $10 Billion 'Licensing Fee' to the Bermuda branch."

Poof. The US profit is now zero. The US tax bill is zero. And Bermuda has a 0% tax rate.

This is perfectly legal. It is the standard operating procedure of the S&P 500.

You cannot fight this with more laws. You cannot fight this with more auditors. As long as "Value" is defined by "Dollars," and Dollars can be moved at the speed of light, the corporations will always win.

We need to switch to **Physical Accounting**.

We don't tax the dollars. We tax the **Physics**.

There are two things that AI cannot fake. There are two things that cannot be routed through a shell company in the Cayman Islands. There are two things that obey the laws of Thermodynamics, not the laws of the Cayman Islands.

1. **Watts** (Energy consumed).
2. **FLOPS** (Floating Point Operations per Second - The raw calculation).

Every Data Center is a factory. It takes in electricity and it puts out Intelligence. It is a physical machine that exists in physical space.

It consumes vast amounts of power. It generates vast amounts of heat.

This is the Achilles' heel of the digital economy. It is tethered to the grid.

Section 3: The Digital Tollbooth

Under the Dividend Mandate, every commercial Data Center in the United States—every server farm, every Bitcoin mine, every AI training cluster—will have a "Smart Meter" installed on the main power line.

It is a **Digital Tollbooth**.

The logic is brutally simple: **If the machine does the work of a human, the machine should pay the tax of a human.**

Here is how it works:

Scenario A: Training a Model (The "Education" Phase)

You are OpenAI. You want to train GPT-6. This requires 25,000 Nvidia GPUs running at 100% capacity for six months.

This process consumes 50 Gigawatt-hours of energy—enough to power a small city.

The meter spins. You pay a surcharge on every kilowatt-hour. This is your "Tuition Tax."

Scenario B: Running an Agent (The "Labor" Phase)

You are a bank. You fire your 5,000 customer service reps and replace them with an AI Agent.

Every time a customer asks a question, the Agent runs an inference calculation. This burns a fraction of a cent in electricity and hardware wear.

The meter spins. You pay a micro-tax on every inference. This is your "Payroll Tax."

Scenario C: High-Frequency Trading (The "Speculation" Phase)

You are a hedge fund. You use an AI to execute 1,000 trades per second to scalp pennies from the market.

You are burning compute to extract value.

The meter spins. You pay the toll.

This revenue stream flows directly into the US Treasury to fund the government, replacing the Income Tax.

Why is this the perfect tax?

1. It is Unavoidable.

You can hide a bank account. You can hide a patent. You can hide a transaction.

You cannot hide a 500,000-square-foot warehouse filled with burning-hot silicon chips.

Data Centers are massive physical objects. They require power lines thick enough to kill an elephant. They require millions of gallons of water for cooling.

They are the easiest things in the world to audit.

2. It Scales with the Economy.

As AI gets more powerful, it uses more compute. (Moore's Law).

As the economy becomes more automated, the tax base grows automatically.

We don't need to pass new laws every year. The revenue scales with the technology.

3. It Incentivizes Efficiency.

If companies are taxed on their energy/compute usage, they will race to make their AI more efficient. They will try to get more intelligence for less power.

This is good for the environment and good for the grid.

Section 4: The Eye of Sauron (The Unavoidable Audit)

But what if they cheat? What if they try to run "Secret AI" off the grid? What if they try to hide the servers in a basement?

We have the technology to catch them. We have **The Eye of Sauron**.

We use satellites equipped with **Thermal Imaging**.

A GPU running at full capacity reaches temperatures of 80°C to 90°C. A data center is essentially a massive heater. It vents hot air into the atmosphere.

From space, a secret data center looks like a volcano.

We can scan the continental United States in real-time. If we see a massive heat bloom in the middle of the Nevada desert that doesn't match a registered facility, we know someone is training a model off the books.

We send the drones. We shut it down.

Physics is the only auditor that doesn't take bribes.

Furthermore, we control the Grid.

You cannot generate the megawatts required to train a frontier model using a diesel generator in your backyard. You need the High-Voltage Grid.

The Utility Companies know exactly where the power is going. By nationalizing the data from the Utility Companies, we create a transparent ledger of where the "work" is being done.

There is no "Offshore" for electricity. The electrons are consumed here. The tax is paid here.

Section 5: The Moral Shift (Blood vs. Electricity)

The economic argument for the Compute Tax is strong. But the **Moral Argument** is stronger.

Why do we tax humans?

We tax humans because we assume that "Work" is something humans do. We assume that labor is the source of value.

But think about the cruelty of the Income Tax. You are taxing a biological organism that is trying to survive.

- You are taxing the nurse who is on her feet for 12 hours, destroying her joints.
- You are taxing the teacher who is buying supplies out of her own pocket.
- You are taxing the construction worker risking his life on a scaffold.

You are taxing Sweat.

You are taxing Blood.

You are taxing Time.

Sweat, blood, and time are precious. They are finite. A human only has so many heartbeats. To tax a human's effort is to tax their very life force. It is a penalty for trying to stay alive.

A machine feels no pain. A GPU does not get tired. A server does not have a family to feed. It does not have anxiety. It does not get depressed.

The Compute Tax is the moral shift from taxing **Biological Struggle** to taxing **Electronic Friction**.

Universal Truth: Tax the Watts, not the Sweat.

Let the machines pay for the civilization they are disrupting.

If an AI company wants to use the American energy grid, the American legal system, and the American internet to build a God-Mind that generates trillions in value, they can pay a surcharge on the electricity that powers it.

This is the end of the "freeloader" economy where Tech Giants extract billions while paying nothing.

Under Dividendism, the machine pays for the road it drives on.

Section 6: The Ecosystem of Freedom

So, we have built the machine. Let's step back and look at

the blueprint of the Solution. The Dividend Mandate is not a patchwork of policies. It is a closed-loop ecosystem designed for the Post-Labor Age.

It consists of two massive pumps:

Pump 1: The Sovereign Fund (The Equity)

- **Source:** 20% ownership of all corporate equity (The "Grand Swap").
- **Destination:** The People.
- **Purpose:** To provide the **Dividend**. This is your "Freedom Money." This replaces the wage. It connects you to the upside of the economy. It creates customers. It solves the Demand Problem.

Pump 2: The Compute Tax (The Revenue)

- **Source:** A tax on Watts and FLOPS (Physical constraints).
- **Destination:** The State (Government).
- **Purpose:** To fund the **Infrastructure**. This pays for the roads, the military, the courts, and the healthcare system. It replaces the Income Tax. It ensures the government remains funded even when human jobs disappear.

The Result:

The Human is Decoupled from the need to survive.

You no longer work to eat. You eat because you are a shareholder.

Does this mean nobody works?

No. It means nobody toils.

We will still have work. But it will be the work of humans, not robots. We will have nurses, artists, philosophers, explorers, parents, community builders. We will have people doing things because they matter, not because they are terrified of starvation.

We free the human spirit from the Hamster Wheel.

- The Robots do the heavy lifting.

- The Robots pay the bills.
- The Robots generate the wealth.
- And we—the Masters, the Creators, the Shareholders—we live.

This is not a fantasy. The math works. The technology is here. The Game Theory (Door #3) proves it is the only way to avoid collapse.

The only thing missing is the **Will**.

The only thing standing between you and this future is the belief that the old rules are sacred. They are not sacred. They are just software. And it is time for an update.

But how do we install this update?

The Elites will not give this to us. They will not hand over 20% of their empire because we asked nicely in a book. They will only do it if they are forced.

We have the Theory. We have the Solution. Now, we need the Action.

How do you survive the transition? How do you protect yourself while the old world is burning but the new world hasn't been born yet? How do you navigate The Great Turbulence of the next five years?

Part IV is your survival guide.

It's time to talk about what you need to do tomorrow morning.

SYSTEM LOG: PATCH INSTALLED

STATUS: SOLUTION ARCHITECTURE DEPLOYED

SUBJECT: DIVIDENDISM (CAPITALISM 2.0)

SUMMARY:

- **The Three Doors:** We face three futures: The Bunker (Elysium), The Zoo (UBI), or The Shareholder State (Dividendism). Only the third option preserves human dignity.
- **The Grand Swap:** Corporations trade 20% Equity for 0% Tax. This aligns the incentives of the AI (Growth) with the incentives of the People (Survival).
- **The Sovereign Fund:** We treat the nation like a corporation. Every citizen is a shareholder. The dividend is not welfare; it is a royalty check for our data.
- **The Compute Tax:** We stop taxing human sweat (Income Tax) and start taxing machine friction (Watts & FLOPS). The machine pays for the civilization it disrupts.

ACTION REQUIRED: PROCEED TO PART IV FOR INDIVIDUAL SURVIVAL PROTOCOLS.

PART IV: THE ACTION - THE OWNERSHIP REVOLUTION

CHAPTER 10: THE GREAT TURBULENCE

Section 1: The Gap

Welcome to the most dangerous five years of your life.

We know that the Wage System is dead. We know that the Zero Theory makes abundance inevitable. We know that Dividendism is the only mathematical solution to avoid collapse.

But there is a catch. A terrifying, potentially fatal catch.

Ideas move at the speed of light. Politics moves at the speed of a sloth. The technology to replace you is already here. The laws to protect you are not.

We are entering **The Gap**.

Imagine a rope bridge suspended over a canyon.

Behind you, the cliff is on fire. That is the "Job Market" of the 20th Century. It is burning down. The flames of automation are licking at the ropes. You cannot go back. You cannot "return to normal."

Ahead of you, on the other side of the canyon, is the solid ground of the "Dividend Era." It is the lush, green future where the machines work and the humans live. It is the destination we described in Part III.

But the bridge doesn't reach the other side yet. The planks end in mid-air.

We are walking out over the abyss, hoping that the engineers (politicians) build the rest of the bridge before we run out of footing.

This is the period between **2026 and 2030.** It is the transition zone. It is the time when the Old System is dead, but the New System has not yet been born.

In this Gap, the rules of reality will break down. You are going to witness **Economic Schizophrenia**.

For the next few years, you will look at the news and you will think you are losing your mind. You will see two contradictory headlines on the same screen, and both will be true.

Headline 1: "S&P 500 Hits All-Time High. Dow Jones Clears 60,000. AI Tech Stocks Vertical."

Headline 2: "Homelessness Encampments Double. Food Bank Lines Stretch for Miles. Unemployment at 15%."

How is this possible?

Because in The Gap, the **Economy** and **Society** completely decouple.

In the old world, a booming stock market meant companies were hiring. It meant people had money. It meant the "consumer" was strong.

In the AI world, a booming stock market means companies are **firing**.

Every time a major corporation announces a layoff of 10,000 workers to replace them with an AI agent, their stock price will jump 5%. Wall Street loves efficiency. Wall Street loves margin. And there is nothing more efficient than a workforce that requires zero salary.

So, we will see a **Melt-Up** at the top. The trillions of dollars in productivity gains from AI will flow instantly into the hands of the shareholders (The Top Jaw of the Crocodile). The rich will get richer at a speed that defies physics. They will be buying islands while you are clipping coupons.

Simultaneously, we will see a **Melt-Down** at the bottom.

The demand destruction caused by mass layoffs will crush local businesses. The restaurant, the barber shop, the local accountant—they will starve because their customers have no money.

This is the eye of the storm.

Section 2: The Lie of Statistics (Phantom Jobs)

And the government? They will lie to you. They will lie with statistics. They will try to hide the body.

They cannot admit that unemployment is hitting 15% or 20%. That would cause a panic. That would cause a revolution. That would cause them to lose the next election.

So, they will redefine the word "Employed."

Welcome to the era of **Phantom Jobs**.

If you lose your $80,000-a-year career as a graphic designer and start driving Uber for 10 hours a week to survive, the Department of Labor calls you "Employed."

If you do three hours of "freelance consulting" on Upwork for $50, the government ticks the box. "Employed."

If you are selling your furniture on eBay to pay rent, you are an "Entrepreneur."

This is the **Gig Camouflage**.

They will flood the zone with buzzwords. "The Flex Economy." "The Creator Economy." "Micro-Entrepreneurship."

Do not be fooled. These are euphemisms for **Digital Day Labor**.

They are masking the collapse of the Full-Time Job (with benefits, stability, and dignity) by counting every desperate grasp for cash as a "Job."

They will tell you the unemployment rate is a healthy 4%. But you will feel the truth. You will see it in your neighborhood. You will see it in the "For Lease" signs on every storefront. You will see it in the anxiety in your friends' eyes.

You are not crazy. The data is rigged.

This is the danger of The Gap. It is a period of maximum vulnerability.

Section 3: The Acceleration

The corporations know that Dividendism is coming. They know that eventually, the government will force them to pay. They know the free ride cannot last forever.

So, what will they do in the meantime?

They will **sprint**.

They will accelerate the automation. They will try to extract as much wealth as possible *before* the new laws are passed. They will try to build their moats and secure their monopolies while the chaos distracts the population.

They will try to lock in the "Techno-Feudal" structures we discussed in Chapter 3. They will tighten the Terms of Service. They will raise the subscription prices. They will lay off the humans as fast as the servers can spin up.

It is a looting operation. They are stripping the copper wiring out of the walls of the Middle Class house before the sheriff arrives.

And you?

You are on your own.

This is the hardest pill to swallow. I have outlined the Solution in Part III. I believe we will get there. I believe the Dividend is inevitable.

But it might take five years. It might take ten.

History is messy. The "Three Doors" negotiation won't happen over a cup of tea. It will happen in the streets. It will happen after banks fail. It will happen after elections are fought and lost.

If you sit on your couch waiting for the "American Sovereign Fund" to send you a check tomorrow, you will starve.

You cannot wait for the savior. The government is a slow, reactive beast. It only fixes potholes after someone has already broken an axle. It will not fix the economy until the riots are on the White House lawn.

You must survive **The Gap** yourself.

Section 4: The Time of Monsters

The Italian philosopher Antonio Gramsci wrote from a prison cell in the 1930s:

"The old world is dying, and the new world struggles to be born: now is the time of monsters."

We are in the time of monsters.

- The monsters are the algorithms that fire you via email.
- The monsters are the landlords raising rent by 20% while wages stay flat.
- The monsters are the politicians telling you "the economy is strong" while you are drowning.
- The monsters are the scammers using deepfake voices to steal your grandmother's savings.

To survive the monsters, you cannot be a sheep. You must become a wolf.

You must ruthlessly cut your dependency on the wage system. You must aggressively acquire the only things that hold value. You must harden your mind against the gaslighting of the media.

The Gap is a test. It is the fire that burns away the dead wood of the 20th century. It is going to be scary. It is going to be unfair. It is going to be volatile.

But on the other side of The Gap lies the Dividend. On the other side lies the Golden Age.

Your only mission—your *only* job for the next five years—is to keep your footing on the bridge. Do not look down. Do not stop moving.

So, what do you put in your backpack for the crossing? What assets survive the Melt-Up and the Melt-Down? If money is broken, and jobs are gone, what is "Safe"?

In the next section, we open the survival kit. We talk about **Asset Sovereignty**.

We talk about the only investments that the Black Hole cannot eat.

Universal Truth: The old world is dying, and the new world struggles to be born. Now is the time of monsters.

CHAPTER 11: THE DOCTRINE OF ASSET SOVEREIGNTY

Section 1: The Saver's Funeral

Your grandfather gave you advice that is going to kill you.

He told you: "Work hard. Spend less than you earn. Put the rest in a savings account. Compound interest will make you rich."

In 1950, this was true. In 2026, this is a suicide pact.

We established in Chapter 10 that during "The Gap," the government will likely choose Door #2 (Money Printing) to prevent riots before they are forced into Door #3 (The Dividend). They will print trillions to fund stimulus checks, bailouts, and "soft" UBI.

When the supply of money goes to infinity, the value of money goes to zero.

If you are holding cash—dollars, euros, yen—you are holding a melting ice cube. You are working for a currency that is being debased faster than you can earn it.

The inflation rate is not the CPI number the government publishes (3% or 4%). That is a manipulated statistic. The *real* inflation rate is the rate at which the money supply expands. If the Fed prints 15% more dollars this year, your savings just lost 15% of their purchasing power.

If your bank pays you 0.5% interest, and the real inflation is 15%, you are losing 14.5% of your wealth every single year.

You are not saving. You are bleeding.

The first rule of the Dividend Mandate is **The Death of Savings**.

You must stop viewing "Money" as a store of value. Money is a transactional medium. It is for buying milk and paying rent. It is not for wealth preservation.

To survive The Gap, you must convert your labor into something that cannot be printed. You must move from being a **Renter of Currency** to an **Owner of Assets**.

This is **Asset Sovereignty**.

It is the declaration that you will no longer store your life force in the paper tickets issued by a bankrupt government. You will store it in reality.

Section 2: The Barbell Strategy

How do you position yourself? The economy is schizophrenic. It is melting up (stocks) and melting down (jobs) at the same time.

If you go 100% into safety (Gold/Bonds), you miss the AI boom and get left behind.

If you go 100% into risk (Tech Stocks/Crypto), a market crash wipes you out.

You need a strategy that handles both extremes. You need **The Barbell**.

Imagine a weightlifting barbell.

- On one side, you have heavy weights. This is **The Shield**. (Safety/Defense).
- On the other side, you have heavy weights. This is **The Spear**. (Growth/Offense).
- In the middle, the bar is empty. You hold **zero** in the middle. The "Middle" is cash, bonds, and mediocre companies.

The Strategy:

You put 50% of your net worth into assets that protect you from inflation (The Shield).

You put 50% of your net worth into assets that profit from the AI revolution (The Spear).

You ignore everything else.

Section 3: The Shield (Hard Assets)

The purpose of The Shield is not to make you rich. It is to keep you from becoming poor. It is to ensure that no matter how much money the politicians print, your purchasing power remains constant.

The Shield is made of **Atoms**, not Bits.

1. Gold (The King):

Gold is not an investment. It is insurance. It is the only financial asset that is not someone else's liability.

If you own a bond, the government owes you. If you own a stock, the company owes you.

If you own a gold coin, nobody owes you anything. You own the metal.

For 5,000 years, Gold has held its value. In Roman times, an ounce of gold bought a good toga and a pair of sandals. Today, an ounce of gold buys a good suit and a pair of shoes.

It is the anchor.

In a world of digital abundance (Zero Marginal Cost), the only thing that retains value is physical scarcity. You cannot "print" gold. You have to mine it.

2. Land (The Fortress):

They aren't making any more of it.

If you can own your home, you have hedged your biggest

expense (rent). If you can own productive land (farmland), you have hedged your survival (food).

Real Estate is a hard asset. It tracks inflation. When the dollar collapses, the price of the house goes up.

3. Potential Energy (The Universal Constant)

This is the ultimate truth of the physical universe: **You cannot print a kilowatt.**

Money is, at its core, just a claim on energy. It is a token that represents work done or work to be done. When the government prints trillion of dollars without creating new energy, they are simply diluting your claim on reality.

The Laws of Thermodynamics do not care about the Federal Reserve. Energy cannot be created or destroyed, only transferred.

In the AI Era, Energy is the bottleneck. The Data Centers that power the "Silicon God" are ravenous beasts. They consume electricity like civilizations consume water.

- To run the AI, you need **Electricity** (Uranium, Natural Gas, Solar).
- To build the grid, you need **Metals** (Copper, Lithium, Nickel).

These are **Hard Assets**. They are "Potential Energy" stored in the ground.

While the dollar loses value, the cost of extracting a barrel of oil or mining a pound of copper stays anchored to physics. By owning commodities—or the companies that extract them—you are hedging against the debasement of paper money.

If the economy collapses, people still need heat. If the economy booms (AI takeover), the machines need power. Energy wins in both scenarios. It is the only currency that never goes to zero.

Action Item: Look at your savings. If it is sitting in a checking account, you are naked. Move it into **The Shield**.

1. Buy **Physical Gold** (The Anchor).
2. Buy **Land/Real Estate** (The Fortress) if you can.
3. Buy **Energy/Commodities** (The Fuel). If you cannot buy an oil well, buy the ETFs that track energy and critical metals. Get out of the Dollar. Get into Physics.

Section 4: The Spear (Owning the Robots)

The Shield keeps you alive. The Spear makes you free.

The Shield protects you from the government (Inflation). The Spear allows you to profit from the Corporation (AI).

Recall the "Grand Swap" from Chapter 8. We want the government to own 20% of the AI companies. But until that happens, **you** must own them yourself.

If AI is going to steal your job, you must own the AI.

This is the concept of the **Personal Hedge**.

If you are a truck driver, and self-driving trucks are coming, you are "Short" automation. If automation happens, you lose.

To balance this, you must go "Long" automation. You must buy stock in Tesla or Waymo.

If the robots take your job, the stock goes up. The profit from the stock offsets the loss of the wage.

What to Buy:

Do not try to pick the "Winner." You will fail. You don't know if Google or OpenAI or Anthropic will win the war.

Buy the Infrastructure.

Buy the "Shovel Sellers" we discussed in Chapter 5.

- **The Chip Makers:** (Nvidia, AMD, TSMC). They make the brains.
- **The Cloud Lords:** (Microsoft, Amazon, Google). They own

the servers.
- **The Energy Grid:** (Nuclear, Utilities). They power the beast.

Or, simpler: Buy the Index (S&P 500 or Nasdaq 100).

The S&P 500 is effectively an AI fund now. The top 7 companies drive all the returns. By owning the index, you own the revolution.

The Math of the Spear:

The AI economy will generate trillions in wealth. That wealth will not go to wages. It will go to Earnings Per Share (EPS).

If you own the shares, you get the wealth.

If you do not own the shares, you get nothing.

It is that binary.

You must be an **Owner**. Even if you only have $100. Buy a fractional share. Start building the Spear.

Section 5: The Middle is Death

The danger zone is the middle.
- **Bonds:** Bonds are a promise to pay you back in dollars. Why would you want more dollars in an inflation crisis? Bonds are "Return-Free Risk."
- **Mid-Cap Companies:** Small, inefficient companies that rely on labor will be crushed by the AI giants. They are the "Kill Zone."
- **Cash:** We already covered this. Cash is trash.

Stick to the edges.

Maximum Safety (Gold).

Maximum Growth (Tech).

This is the Barbell. It is resilient.

If the economy crashes, the Gold saves you.

If the economy booms (inflationary melt-up), the Tech stocks make you rich.

If the dollar dies, the Gold and the Assets survive.

Section 6: The Mental Shift

This requires a psychological rewiring.

You have been trained to be a consumer. You get $100, you buy a pair of sneakers.

You must train yourself to be a capitalist. You get $100, you buy a share of Nike.

Every dollar you spend on a "thing" is a vote for your own poverty.

Every dollar you spend on an "asset" is a brick in your fortress.

During The Gap, you must live like a monk.

Cut your expenses. Stop buying plastic junk. Stop subscribing to services you don't use.

Channel every available ounce of financial energy into The Shield and The Spear.

You are building your own personal Sovereign Wealth Fund.

You are building your own personal Alaska Permanent Fund.

You cannot wait for the government to pass the Dividend Mandate. You must pay the dividend to yourself first.

Universal Truth: You cannot work your way out of a deflationary collapse. You can only invest your way out.

The window is closing. The prices of these assets are rising. The "Melt-Up" has begun.

The rich are front-running you. They are buying the gold. They are buying the Nvidia stock. They are buying the farmland.

Do not be left holding the paper bag.

Get your assets. Secure your sovereignty.

And once you have secured your financial survival, you must secure your mind. Because the war isn't just economic. It is psychological.

The next chapter is about how to keep your sanity in a world of infinite fake noise.

CHAPTER 12: THE HUMAN FORTRESS

Section 1: The War for Your Attention

You are under attack.

You are not being attacked by a foreign army. You are not being attacked by a virus. You are being attacked by a super-computer located in a server farm in Palo Alto that is running a trillion-dollar algorithm designed to hack your dopamine receptors.

The enemy is not "Technology." The enemy is **Extraction**.

Just as they extract your data to train their models (Chapter 4), they extract your **Attention** to sell their ads. They have weaponized psychology. They have hired the best neuroscientists in the world to figure out exactly what color of red notification badge makes you click. They know exactly how many seconds of a video you will watch before you get bored. They know your insecurities better than your spouse does.

They have turned your phone into a slot machine, and you are the addict pulling the lever.

This is the **Algorithmic Psy-Op**.

If you want to survive the AI age, you must first reclaim your mind. You cannot be a "Sovereign Shareholder" if you are a "Digital Junkie."

You must build a **Human Fortress**.

This starts with a declaration of **Mental Sovereignty**.

You must decide that your attention is the most valuable asset you own. It is the spotlight of your consciousness. Wherever you point it, reality exists. If you point it at a screen for 10 hours a day, watching 15-second clips of people dancing or arguing, you are deleting your own reality.

You are living in *their* simulation, not your life.

Section 2: The Analog Protocol (Monk Mode)

How do you fight a super-computer? You don't. You unplug it.

The most radical act of rebellion in 2026 is to be **Offline**.

I am not telling you to become a Luddite and smash your phone. The phone is a tool. But it must be a tool, not a master.

You need to implement the **Analog Protocol**.

1. The Phone Foyer Rule:

When you walk into your house, the phone goes in a basket by the door. It stays there. It does not go to the dinner table. It does not go to the bedroom.

The bedroom is a sanctuary. No screens. No blue light. Only sleep and intimacy.

If you wake up and the first thing you do is scroll, you have lost the battle before the day has begun. You have let the world dictate your mood.

2. The Deep Work Block:

The AI is fast, but it is shallow. It cannot do "Deep Work" (as Cal Newport defined it). It cannot hold a complex philosophical thought for four hours. It hallucinates.

To compete with the machine, you must go where the machine cannot. You must go Deep.

Carve out 4 hours a day. No internet. No wifi. Just you and the work. Write, code, build, think.

In a world of distracted squirrels, the focused monk is King.

3. The Physical Check-In:

Every day, you must touch grass. Literally.

You must feel the sun. You must lift a heavy weight. You must sweat.

Your body is the hardware that runs your mind. If the hardware is weak, the software glitches.

The Metaverse wants you to be a floating head in a jar. Resist. Be an animal.

Section 3: The High-Touch Sanctuary

We talked about the "Singularity of Value" in Chapter 6. We said that as "Digital" becomes free, "Physical" becomes expensive.

This applies to your skills, too.

If your job involves sitting at a computer and moving information from Column A to Column B, you are in the Kill Zone. The AI will do that faster and cheaper.

You must migrate to the **High-Touch Economy**.

These are the skills that require **Empathy**, **Leadership**, and **Physical Reality**.

1. The Leader:

AI can generate a plan. It cannot inspire a team to execute the plan. It cannot look a frightened employee in the eye and give them courage. It cannot navigate the messy, irrational politics of a boardroom.

Leadership is about Trust. Humans only trust humans.

2. The Craftsman:

AI can design a chair. It cannot build the chair.

The plumber, the electrician, the carpenter, the nurse, the

surgeon, the chef. These jobs require interaction with the chaotic physical world.

A robot hand is still clumsy. It cannot feel the tension in a screw. It cannot feel the heat of a fever.

If you work with your hands, you are safe for a long time.

3. The Synthesizer:

AI is great at analysis (breaking things down). It is bad at synthesis (putting things together in a new way).

The AI can write a legal brief. It cannot look at the client, understand their moral dilemma, look at the political climate, and craft a strategy that wins the case outside the courtroom.

Be the architect, not the bricklayer.

Action Item:

Audit your career.

Ask yourself: "Can a software update do my job?"

If the answer is "Yes," pivot immediately.

Pivot to people. Pivot to the physical. Pivot to the complex.

Section 4: The Return to the Tribe

The greatest lie of the digital age is that we are "connected."

We are more isolated than ever. We have 5,000 "friends" on Facebook and nobody to help us move a couch.

Loneliness is a national security threat. A lonely population is a vulnerable population. Lonely people join cults. Lonely people get addicted. Lonely people give up.

To build the Human Fortress, you must rebuild the **Tribe**.

You need a "Third Place."

The First Place is Home. The Second Place is Work. The Third Place is where you exist as a citizen.

The church, the gym, the book club, the softball league, the pub.

These places are dying. You must resurrect them.

You need to cultivate **Real-World Networks**.

If the internet goes down tomorrow—if the "Cyber-Polygon" event happens—who do you know?

Do you know your neighbors? Do you know a guy who can fix a generator? Do you know a doctor who lives on your street?

These relationships are your true safety net.

In the "Great Turbulence" (Chapter 10), the government might fail you. The supply chain might break.

Your neighbor won't fail you—if you have built the bond.

The Dinner Party Rule:

Host a dinner party once a month. Cook food. Invite people. No phones allowed at the table. Talk.

It sounds trivial. It is revolutionary.

You are weaving the fabric of society back together, one meal at a time.

Section 5: The Optimism of Agency

Finally, the Human Fortress requires a specific mindset: **Tragic Optimism**.

This is not "Toxic Positivity" (everything is great!).

This is the belief that: "Life is hard, suffering is inevitable, but I have the agency to make it meaningful."

The AI Doomers will tell you it's over. They will tell you the robots will kill us all.

The Technocrats will tell you to just surrender and take the UBI.

Reject both.

Believing in the future is a self-fulfilling prophecy.

If you believe you are obsolete, you will become obsolete.

If you believe you are the master of the machine, you will master it.

We are the species that discovered fire. We are the species that built the pyramids. We are the species that went to the moon.

We are not going to be replaced by a chatbot unless we choose to be.

The Human Spirit is the only thing in the universe that can create Meaning.

The AI can generate text. It cannot generate Meaning.

That is your monopoly.

Hold the line.

Keep your mind sharp. Keep your body strong. Keep your family close.

And when the check arrives—when the Dividend is finally paid—you will be ready to use it not just to survive, but to Thrive.

This concludes **Part IV: The Action**.

It is time to paint the picture of the destination. It is time to show what the world looks like after we win.

SYSTEM LOG: USER UPGRADE

STATUS: ROOT ACCESS GRANTED

SUBJECT: THE OWNERSHIP REVOLUTION

SUMMARY:

- **The Gap (2026-2030):** The transition period will be chaotic. The old world is dying; the new is not yet born. You must survive the turbulence.

- **The Barbell Strategy:** Hedge your survival. 50% in The Shield (Gold/Land) to protect against inflation. 50% in The Spear (AI Equity) to profit from the boom.
- **Asset Sovereignty:** You cannot work your way out of a deflationary collapse. You can only invest your way out. Own the robots.
- **The Human Fortress:** Protect your mind from the algorithmic psy-op. Disconnect to reconnect. Build high-touch, real-world tribes.

ACTION REQUIRED: PROCEED TO PART V FOR THREAT ASSESSMENT.

PART V: THE COUNTER-REFORMATION

CHAPTER 13: THE IMMUNE RESPONSE (THE EMPIRE STRIKES BACK)

Section 1: The Biological Imperative of the State

If you treat the Global Economy as a machine, you will misunderstand it. Machines are static. Machines break and stay broken until fixed.

You must treat the System—the interlocking nexus of Central Banks, Federal Governments, and Mega-Corporations—as a **biological organism**.

A biological organism has one primary directive: Survival.

It has one primary defense mechanism: The Immune System.

In Part IV, I handed you the blueprints for sovereignty. I told you to buy **Gold and Energy (The Shield)**. I told you to acquire equity in the AI giants (The Spear). I told you to build the Human Fortress. I told you to exit the dependency loop.

Did you think the System would just let you leave?

Did you think the Federal Reserve, the IRS, and the monopolies of Silicon Valley would watch you detach from their power grid and simply wave goodbye?

No.

When a biological organism detects a foreign body—a virus, a bacteria, or a rogue cell that refuses to obey the central DNA—it launches an immune response. It sends white blood cells to surround, suffocate, and dissolve the intruder.

You are the intruder.

The rise of Crypto, the democratization of AI, and the concept of Dividendism represent an existential threat to the Old Order.

- **Crypto** threatens the State's monopoly on money.
- **Open Source AI** threatens the Corporation's monopoly on intelligence.
- **Dividendism** threatens the Elite's monopoly on capital accumulation.

We are currently witnessing the beginning of the **Great Counter-Reformation**.

Just as the Catholic Church launched a brutal counter-offensive against the Protestant Reformation to maintain its grip on the souls of Europe, the Techno-Statist Alliance is launching a counter-offensive to maintain its grip on the wallets of America.

They will not send soldiers to your door. That is too messy. That is too kinetic.

They will use silent weapons. They will use code, law, and inflation.

They are building a digital cage, and they are painting it gold so you will walk into it willingly.

This chapter is your intelligence briefing on the enemy's battle plan. You need to know what is coming.

Section 2: The CBDC Trap (The Panopticon of Money)

The greatest trick the devil ever pulled was convincing the world that Central Bank Digital Currencies (CBDCs) are "Crypto."

They will sell it to you as an upgrade. They will say: "Look! It's like Bitcoin, but faster! It's safer! It's backed by the full faith and credit of the United States Government!"

Do not listen.

CBDCs are the anti-matter of Crypto.

Bitcoin is decentralized, permissionless, and private. It is money that answers to no master.

A CBDC is centralized, permissioned, and surveilled. It is money that answers only to the master.

To understand the danger, you must understand the difference between **Money** and **Vouchers**.

- **Money** is a bearer instrument. If you have a $20 bill, you own the value. You can spend it on anything, anywhere, at any time. The transaction is between you and the merchant.
- **A Voucher** (like a gift card or a food stamp) is a conditional instrument. It is valid only *if* the issuer says it is.

A CBDC converts the US Dollar from Money into a Voucher.

It is **Programmable Money**. This means the Federal Reserve can write code into the currency itself that dictates how, when, and where it can be used.

Consider the "Stimulus Check" scenario.

In 2020, the government mailed checks. You could save that money, invest it, or spend it.

With a CBDC, they can issue a "Stimulus Token" with an Expiration Date.

"This $1,000 must be spent within 30 days, or it dissolves."

Why? To force velocity of money. To prevent you from saving. To force you to consume to prop up their GDP metrics.

Consider the "Climate Lockdown" scenario.

They can attach a Carbon Limit to your digital wallet.

"You have purchased your maximum allotment of gasoline and red meat for this month. Your wallet is now declined for these categories until the 1st of next month."

Consider the "Political Dissident" scenario.

We saw a preview of this in Canada during the Trucker Protests. The government froze the bank accounts of protesters without due process.

With a CBDC, they don't need to call the bank. They are the bank. They can toggle your ability to participate in the economy from "On" to "Off" with a single keystroke.

This is the Financial Panopticon.

Every transaction you make—every coffee, every book, every donation—is recorded on a centralized ledger visible to the IRS, the FBI, and the NSA. There is no privacy. There is no "under the table." There is no escape.

They are building this right now. The "FedNow" payment rails are the infrastructure. The pilot programs are running.

Their goal is to prevent Capital Flight.

They know the Dollar is dying (we will get to that in Section 3). They know you want to move your wealth into The Shield (Gold and Energy).

A CBDC allows them to build a digital Berlin Wall. They can make it impossible to convert "FedCoins" into Gold. They can trap you in the burning building.

The Counter-Move:

This is why you must build your Shield now. You must be positioned in hard assets before the gates close. Once the currency is fully digitized, the exit ramps will be demolished.

Section 3: The Regulatory Moat

(The Drawbridge Strategy)

While the State comes for your money, the Corporations are coming for your tools.

We discussed "The Spear"—owning the AI revolution. But the tech giants (OpenAI, Google, Microsoft, Anthropic) have a problem.

They are terrified of Open Source.

In 2023, a leaked Google memo stated plainly: "We have no moat, and neither does OpenAI."

They realized that a kid in a garage with a laptop could download an open-source model (like LLaMA), fine-tune it, and run it locally for free.

If AI becomes free and abundant, their profit margins collapse.

So, they are executing **The Drawbridge Strategy**.

They are lobbying governments worldwide for "AI Safety" regulation.

They are flying to Washington, London, and Brussels. They are testifying before Congress with solemn faces. They are saying:

"AI is dangerous. It could build bioweapons. It could destroy democracy. You must regulate it. You must require licenses for training large models."

This sounds like responsibility. It is actually **Regulatory Capture**.

They are begging the government to build a wall around the AI industry.

If the government passes a law saying, "Anyone training an AI model must undergo a $10 million safety audit and hold a federal license," who does that hurt?

- It does not hurt Google. They have billions. They can pay the auditors. They have armies of lawyers.

- It destroys the startup. It destroys the open-source community. It destroys the kid in the garage.

They are pulling up the drawbridge after they have already crossed into the castle.

They want to turn AI into **Nuclear Energy**—a technology that is technically possible, but legally impossible for anyone but a state-sanctioned monopoly to touch.

This is the counter-reformation against decentralized intelligence.

They want you to be a subscriber to their Intelligence, not an owner of your own.

They want to rent you the "God-Mind" for $20 a month, subject to their censorship, their bias, and their terms of service. They do not want you running a God-Mind on your own server.

This connects directly to the "Compute Tax" concept. They will try to monopolize the *hardware* (GPUs) and the *energy* required to run it. They will classify high-end GPUs as "munitions" (which they already have for export to China) and restrict their sale to individuals.

The Counter-Move:

You must support the decentralized AI movement. But more importantly, as an investor, you must understand that the "Magnificent Seven" act as a cartel. They are protected by the State. This makes their stock a safe haven in the short term (The Spear), but a prison in the long term. You must hedge.

Section 4: The Inflationary Solvent (The Silent Theft)

The third prong of the Counter-Reformation is mathematical. It is the **Inflationary Tax**.

The United States government is roughly $35 Trillion in debt (and climbing by $1 Trillion every 100 days).

The interest payments on this debt are now higher than the defense budget.

Mathematically, they cannot pay it back.

A government with unpayable debt has only two choices:

1. **Hard Default:** Admit bankruptcy. Stop paying bonds. This causes a global depression and the collapse of the government. They will never choose this.
2. **Soft Default (Inflation):** Print money to pay the debt.

They will choose Option 2. They always choose Option 2.

Inflation is not an accident. It is policy.

Inflation is a tax that nobody voted for.

When they print 20% more money, they are stealing 20% of your savings. They are transferring purchasing power from Holders of Currency (You) to Holders of Debt (Them).

But here is the "Counter-Reformation" twist:

They know that if inflation gets too high, the people will revolt. So they engage in Financial Repression.

They will manipulate the CPI (Consumer Price Index) to understate real inflation.

They will tell you inflation is 3%, while your grocery bill is up 20%.

They will suppress interest rates so that you cannot earn a yield on your savings.

Their goal is to inflate away the middle class before the middle class can transition to the Dividend Era.

They want to strip you of your assets.

If your savings are melting at 10% a year, and your wages are stagnant, you are running on a treadmill that is speeding up. You are forced to sell your assets just to survive.

Who buys those assets?

The "Cantillon Insiders"—the people closest to the money printer (BlackRock, Banks, Hedge Funds).

They get the fresh money first. They buy the houses, the farmland, and the stocks while you are selling them to buy food.

This is a **Transfer of Wealth** of biblical proportions. It is a vacuum cleaner sucking the equity out of the bottom 90% and blasting it into the accounts of the top 0.1%.

The Counter-Move:

This validates the "Doctrine of Asset Sovereignty" (Chapter 11). You cannot hold dollars. The dollar is the weapon being used against you. You must hold things that cannot be printed. The Inflationary Tax only hits those who hold paper.

Section 5: The Narrative Gaslight (The Happiness Trap)

Finally, the Counter-Reformation requires a psychological component.

You cannot strip people of their wealth without a story that makes them accept it.

Enter the **Narrative Gaslight**.

You have heard the slogan from the World Economic Forum (WEF):

"You will own nothing, and you will be happy."

This is not a conspiracy theory. It is a marketing tagline for the **Subscription Economy**.

The corporate media is currently engaged in a massive psy-op to normalize "Serfdom."

- They tell you that "Renting is better than

owning" (Freedom from maintenance!).
- They tell you that "Van Life is cool" (Homelessness branded as adventure!).
- They tell you to eat bugs and synthetic sludge (To save the planet!).
- They tell you that "insects are a delicacy."

They are rebranding poverty as minimalism.

They are rebranding feudalism as "The Sharing Economy."

Why?

Because the Dividend Mandate requires ownership. It requires you to have equity.

If they can convince you that ownership is "old fashioned," "racist," or "environmentally destructive," they can convince you to voluntarily surrender your claim to the future.

They want a population of "Renters."
- You rent your house from BlackRock.
- You rent your music from Spotify.
- You rent your software from Microsoft.
- You rent your intelligence from OpenAI.

If you stop paying, you stop existing.

This is the ultimate form of control. A renter has no rights. A renter can be evicted. An owner has sovereignty.

The "counter-reformation" is a war on the concept of Private Property itself. They are trying to revert society to a pre-Enlightenment structure where a small Aristocracy owns the land (and the data), and the Peasantry rents the right to exist upon it.

Section 6: The Checkpoint

We have arrived at a dark intersection.

In Part IV, I gave you the map to freedom.

In this chapter, I have shown you the obstacles the Empire is placing on that map.

- **The CBDC** locks the gate to the financial exit.
- **Regulatory Capture** locks the gate to the technological exit.
- **Inflation** burns the bridge behind you.
- **Gaslighting** blinds you to the danger.

They are trying to close the window.

The "Gap" years (2026-2030) are not just a period of economic turbulence. They are a race.

It is a race between The Great Expansion (The Dividend, The Abundance) and The Great Enclosure (The CBDC, The Control Grid).

If the Enclosure happens before the Dividend is secured, we lose. We enter the "Elysium" scenario (Door #1) forever.

This is why you must move with speed.

You cannot wait for the laws to change.

You cannot wait for the "Grand Swap" to be negotiated.

You must secure your own Personal Dividend now.

You must build your Human Fortress now.

You must move your capital into The Shield now.

The Empire is striking back. The immune system is activated.

But the virus—the idea of human liberty backed by automated abundance—is resilient.

They can slow it down. But they cannot stop the math.

Physics always wins in the end.

The only question is: Will you survive long enough to see the victory?

In the next chapter, we will discuss the geopolitical layer. What happens when China, the US, and the Global South realize they are fighting for the last barrel of oil and the first chip of AGI?

SYSTEM LOG: THREAT DETECTED

STATUS: HOSTILE ENTITIES ACTIVE

SUBJECT: THE COUNTER-REFORMATION

SUMMARY:

- **The Immune Response:** The Old Order (Banks/State) will not surrender power quietly. They will launch a counter-offensive to maintain control.
- **The CBDC Trap:** Central Bank Digital Currencies are surveillance tools designed to prevent capital flight. They turn money into vouchers.
- **Regulatory Capture:** Big Tech is lobbying for "AI Safety" laws to build a moat around their monopoly, preventing you from owning your own intelligence.
- **The Inflationary Solvent:** The government will try to inflate away the debt, destroying the middle class's savings before the Dividend arrives.

ACTION REQUIRED: PROCEED TO PART VI FOR GEOPOLITICAL ANALYSIS.

PART VI: THE GEOPOLITICS OF ZERO

CHAPTER 14: THE GEOPOLITICS OF ZERO (THE AI ARMS RACE)

Section 1: The Prisoner's Dilemma of the Species

For the last three decades, globalization was built on a premise of cooperation. We trade; we grow rich; we don't fight. That era ended the moment ChatGPT passed the Turing Test.

We have entered a new era of **Hyper-Realpolitik**.

To understand why the "Dividend Mandate" is inevitable, you must first understand why the "AI Pause" is impossible.

You hear the calls from ethicists and concerned scientists: *"Pause the development of AI! It is too dangerous! We need six months to assess the safety!"*

These people are well-intentioned. They are also functionally illiterate in Game Theory.

The development of Artificial General Intelligence (AGI) is trapped in a global Prisoner's Dilemma.

There are two main players: The United States and China.

Here is the payoff matrix:

1. **If US stops and China stops:** The world is safer (maybe).
2. **If US stops and China builds:** China achieves AGI dominance. They gain a permanent military and economic advantage. The US becomes a vassal state.

3. **If US builds and China stops:** The US achieves AGI dominance. The American Century continues for another millennium.
4. **If both build:** We race to the precipice of the Singularity, risking extinction, but maintaining parity.

Look at the matrix.

Option 1 (Cooperation) requires absolute trust. There is zero trust.

Therefore, the only rational move for both players is Option 4: Race.

There are no brakes on this train. The United States government cannot ban AI development because doing so would be an act of unilateral disarmament. It would be handing the keys to the future to the Chinese Communist Party (CCP).

The "Zero Point" we discussed in Part II—the moment where the cost of intelligence hits zero—is not just an economic event. It is a geopolitical weapon.

Whoever reaches the Zero Point first controls the global operating system.

Universal Truth: You cannot regulate a technology that your enemy is using to build a gun.

Section 2: The Silicon Curtain (The Most Dangerous Place on Earth)

In the 20th Century, wars were fought for Oil.

In the 21st Century, wars are fought for Sand.

Specifically, processed silicon sand, printed with light, packaged in Taiwan.

There is a single building in Hsinchu Science Park, Taiwan, that holds the fate of human civilization. It is the headquarters of TSMC (Taiwan Semiconductor Manufacturing Company).

They are the only entity on earth capable of manufacturing the advanced chips (3nm and below) required to train Frontier Models.

This creates a terrifying geopolitical bottleneck.

- The US designs the chips (Nvidia, Apple).
- The Netherlands builds the machines that print the chips (ASML).
- Taiwan manufactures the chips.
- China wants the chips.

The United States has realized that if China gets access to these chips, they will build an AI that surpasses American capability.

So, in October 2022, the Biden Administration launched the opening salvo of World War III. They didn't fire a missile. They signed an export control order.

They banned the sale of high-end AI chips (Nvidia H100s) to China.

They banned American citizens from working in Chinese chip factories.

They banned the export of the lithography tools.

This is a Technological Blockade. It is the modern equivalent of the oil embargo that triggered Pearl Harbor.

The US is trying to freeze China's AI capability in the year 2022, while the US advances into the 2030s.

This guarantees conflict. China cannot accept permanent inferiority. They are pouring billions into their domestic chip industry (Huawei/SMIC). They are stockpiling chips through black markets in the Middle East.

This creates the Silicon Curtain.

The world is bifurcating into two technology stacks:

1. **The Western Stack:** Nvidia/OpenAI/Microsoft/NATO.

2. **The Eastern Stack:** Huawei/Baidu/CCP/BRICS.

These two stacks will not talk to each other. They will run on different standards, different values, and different truths.

And sitting right on the fault line is Taiwan.

If China blockades Taiwan, the supply of chips stops. The global economy collapses instantly. The "Deflationary Black Hole" turns into an inflationary supernova.

Section 3: Sovereign AI (The Fear of Digital Colonialism)

It is not just the superpowers. Every nation on earth has woken up to a nightmare scenario: **Digital Colonialism.**

Imagine you are France, or Saudi Arabia, or Japan.

You look at the AI landscape. You see that the "Brain" of the world is being trained in California, on American data, with American values, owned by American corporations.

If you simply "subscribe" to ChatGPT, you are importing American culture.

- Your children will learn from an American tutor.
- Your laws will be drafted by an American legal bot.
- Your history will be filtered through an American bias.

You become a "Data Colony." You export your raw data to the US, they refine it into Intelligence, and sell it back to you at a premium.

This is unacceptable to any sovereign nation.

This is why we are seeing the rise of Sovereign AI.

Jensen Huang, the CEO of Nvidia, is traveling the world telling heads of state: *"You cannot outsource your intelligence. You must own the means of producing thought."*

- **Saudi Arabia** is buying thousands of H100 chips to build

"Desert AI."

- **France** is funding Mistral AI to preserve the French language and culture.
- **Japan** is subsidizing domestic chip fabs.

They are treating AI infrastructure not as a business, but as National Defense.

They know that in the future, GDP will not be measured in dollars; it will be measured in Compute.

A nation without Compute is a nation without a future.

This drives the demand for the "Shovels" (Chips/Energy) to infinity.

It also ensures that the "AI Arms Race" will not slow down. It will accelerate. Every nation is terrified of being left behind.

Section 4: The Thucydides Trap 2.0

Historian Graham Allison coined the term "Thucydides Trap."

It refers to the inevitable conflict that occurs when a Rising Power (Sparta/China) threatens to displace a Ruling Power (Athens/USA).

In 12 of the last 16 cases over the past 500 years, this dynamic ended in war.

AI is the accelerant of this trap.

Nuclear weapons created a stalemate (Mutually Assured Destruction) because they are purely destructive. You cannot use a nuke to build an economy.

AI is different. It is Constructive.

If China gets AGI first, they don't just get better weapons. They get:

- Hyper-efficient fusion energy.
- Automated manufacturing that undercuts global prices.

- Biological enhancements for their population.
- Cyber-warfare capabilities that can shut down the US grid without firing a shot.

The winner of the AI race takes everything. It is a "Winner-Take-All" dynamic on a planetary scale.

This creates a period of maximum danger: The Window of Vulnerability.

Right now, the US has a lead (thanks to OpenAI and Nvidia). China is lagging by about 2 years.

China knows this window is closing.

The US knows China is trying to jump through the window.

This implies that the US government will become increasingly desperate to maintain its lead.

They will nationalize the AI industry in all but name.

They will force Google, Microsoft, and OpenAI to work directly with the Pentagon (Project Maven was just the start).

The distinction between "Big Tech" and "The State" will dissolve. They will fuse into a single entity: The US Compute Industrial Complex.

Section 5: The Domestic Front (Why Dividendism is Strategic)

So, how does this relate to **Dividendism**? Why would the US government agree to pay you a dividend?

Because you cannot fight a Cold War with China if your own cities are burning.

Look at the difference between the two systems:

- **China:** An authoritarian system. They can displace workers, automate factories, and suppress dissent with a surveillance state. They can force their population to endure pain for the sake of national dominance.

- **USA:** A democratic(ish) republic. If unemployment hits 20% due to AI automation, and the government does nothing, the US collapses from within. Polarization tears the country apart.

Internal instability is the greatest threat to American Hegemony.

If the US wants to win the AI Arms Race, it needs to unleash the full power of automation. It needs to let the AI rip. It needs to automate the ports, the factories, and the logistics chains to compete with Chinese efficiency.

But if they do that under the current "Wage System," the American working class will revolt. They will vote for populists who promise to "Ban the Bots." They will strike. They will sabotage the grid.

To win the external war, the US government must buy peace on the internal front.

The Dividend is not welfare. It is a bribe.

It is the government saying to the population:

"We need to automate everything to beat China. We need to unleash the Singularity. This will destroy your jobs. But if you let us do this, we will cut you in on the deal. We will give you a share of the Sovereign AI Fund. We will make sure you benefit from the victory."

This is the Roman Grain Dole strategy.

Rome conquered the world, but to keep the peace at home, they gave free grain to the citizens.

It wasn't charity. It was national security.

The US Military and Intelligence agencies are smart. They run simulations. They know that a population that is hungry and unemployed is a liability. A population that is receiving a

"Freedom Dividend" funded by American AI dominance is a loyal asset.

Therefore, the push for Dividendism will not just come from the Left (Socialists) or the Right (Populists).

It will come from the Deep State.

It will come from the Pentagon.

They will realize that **Universal Basic Equity** is the only way to stabilize the biological layer of the American war machine while the silicon layer fights for global dominance.

Section 6: The Inevitability of the Race

We are strapped into a rocket.

The fuse has been lit in Taiwan. The fuel is being poured in Silicon Valley. The target is AGI.

There is no turning back.

The "Luddites" who want to smash the machines will fail because the State cannot allow them to succeed. Smashing the machines means surrendering to China.

The "Regulatory Capture" crowd (Chapter 13) will try to slow it down for *you*, but speed it up for *them*.

Your only move—your only rational move in this Game Theory nightmare—is to align yourself with the vector of the rocket.

- You must own the assets that the State is fighting to protect (Semiconductors, Energy, AI Equity).
- You must position yourself to receive the Dividend when the domestic peace treaty is signed.

The world is breaking into blocks. The Dollar Bloc vs. The BRICS Bloc.

The friction between these tectonic plates will generate heat.

That heat will either burn you, or it will power your transition to the Golden Age.

It depends entirely on where you stand when the silicon curtain falls.

In the next chapter, we look at the final frontier. We look at the philosophical end-game. If we solve the money problem, and we solve the work problem... what is left?

We face the Crisis of Meaning.

SYSTEM LOG: GLOBAL SYNC

STATUS: CONFLICT INEVITABLE

SUBJECT: THE AI ARMS RACE

SUMMARY:

- **The Prisoner's Dilemma:** The US cannot pause AI development because China will not pause. We are locked in a race to the bottom (or top).
- **The Silicon Curtain:** The world is splitting into two technology stacks (US vs. China). Taiwan is the flashpoint.
- **Sovereign AI:** Nations are treating Compute as National Defense. Every country needs its own "Brain" to avoid becoming a digital colony.
- **The Domestic Bribe:** To win the external war against China, the US must buy internal peace. The Dividend is the price of social stability during the transition.

ACTION REQUIRED: PROCEED TO PART VII FOR PSYCHOLOGICAL DIAGNOSTICS.

PART VII: THE MEANING CRISIS

CHAPTER 15: THE CRISIS OF MEANING (THE VOID)

Section 1: The Sunday Neurosis of the Species

Imagine we win.

Imagine the Dividend Mandate is passed. The check hits your account. The AI manages the supply chain. The fusion reactors hum. The cost of living collapses to near zero.

You wake up on a Tuesday morning. You do not have to go to work. You do not have to worry about rent. You do not have to worry about food. The struggle for survival, which has defined every single second of biological evolution for the last 3.8 billion years, is over.

What do you do?

The optimists say we will all become poets and explorers. We will paint masterpieces and debate philosophy.

The pessimists say we will drink ourselves to death and rot in front of VR headsets.

History suggests the pessimists might be right.

Victor Frankl, the psychiatrist who survived Auschwitz, identified a phenomenon he called **"Sunday Neurosis."** He observed that many workers were content during the week, when their time was structured by external demands. But when

Sunday came—the day of rest, the day of freedom—they fell into a deep, existential depression.

Without the external pressure of "The Job," they were forced to confront the emptiness inside them. They realized they had nothing to live *for*.

Now, scale that up. Imagine a civilization that is stuck in a permanent Sunday.

We are about to enter **The Void**.

For 5,000 years, "Meaning" was cheap. It was forced upon us. You had to plant the crop, or you starved. You had to fight the invader, or you died. The meaning of life was simple: **Don't Die**.

But when AI solves the problem of death (or at least, the problem of survival), "Meaning" becomes the scarcest resource in the universe.

We are not the first entities to face this problem. In the 1960s, a scientist named John Calhoun tried to build a paradise. He wanted to see what happens when you give a population everything they want.

The result was not heaven. It was a specific, terrifying kind of hell.

Section 2: The Mouse Prophecy (Universe 25)

John Calhoun built a box. It was a 9-foot square enclosure with high walls. He called it **Universe 25**.

It was a utopia for mice.

- **Unlimited Food:** The hoppers were always full.
- **Unlimited Water:** No drought.
- **No Predators:** No cats, no owls, no traps.
- **Disease Control:** The enclosure was kept clean. Medical care was provided.
- **Perfect Climate:** Always 68°F.

He introduced 4 breeding pairs of mice. Adam and Eve, multiplied by four.

Phase A: The Explosion

At first, it was paradise. The mice realized they were in a land of milk and honey. They bred furiously. The population doubled every 55 days. They were social, active, and vibrant.

Phase B: The Stagnation

But as the population grew, something strange happened. The enclosure could physically hold 3,840 mice. But at around 2,200, the population growth stopped abruptly.

It wasn't a lack of space. There were plenty of empty nesting boxes.

It was a Psychological Collapse.

Because there was no struggle—no need to forage, no need to defend territory from predators, no need to work together—the social fabric disintegrated.

The older mice, who didn't die because there were no predators, refused to give up their roles. The young mice, finding no role to fill, became "Autistic" or "Hyper-Aggressive."

Phase C: The Behavioral Sink

This is where the horror began.

- **The "Outcast" Males:** Young males, unable to find a meaningful role, huddled together in the center of the pit. They became listless during the day and violently aggressive at night. They would attack each other for no reason, biting tails and ears, not for territory, but out of pure, senseless rage.
- **The "Feminist" Females:** The females, abandoned by the males who refused to protect the nests (because there were no threats), became aggressive. They started attacking their own young. They would wound their babies or cast them

out of the nest.

- **The Death of Sex:** Breeding stopped. The males lost the instinct to court. The females lost the instinct to mother.

Phase D: The Beautiful Ones

Then came the final generation. Calhoun called them "The Beautiful Ones."

These were males who withdrew completely from society.

They did not fight.

They did not court females.

They did not mate.

They did nothing but eat, sleep, and groom themselves.

They were physically perfect. Their fur was sleek and unscarred (hence "The Beautiful Ones"). They were healthy.

But they were spiritually dead.

They were "Empty Shells." They had lost the capacity for complex behavior. They were biological robots running a loop of narcissism.

Phase E: Extinction

Eventually, the population crashed. Even when the numbers dropped back down to low levels, the survivors didn't recover. They had forgotten how to be mice. They had forgotten how to bond, how to fight, how to copulate.

The last mouse died in 1973.

Universe 25 went from 0 to 2,200 to 0.

The conclusion was terrifying: **When you remove the struggle for survival, you do not get a "Higher Being." You get a "Beautiful One." You get extinction.**

Section 3: The Modern "Beautiful Ones"

Look around you in 2026. Do you see the mice?

We are seeing the rise of the human "Beautiful Ones."

- **The Hikikomori:** In Japan, millions of young men refuse to leave their bedrooms. They do not work. They do not date. They exist in a digital pod, groomed and fed, but socially dead.
- **The "Lying Flat" Movement:** In China, the youth are "letting it rot." They are opting out of the struggle.
- **The Dopamine Junkies:** In the West, we see young men addicted to video games and pornography. They simulate struggle (in the game) and simulate mating (on the screen), but they engage in neither in reality.

They are safe. They are fed. They have infinite entertainment.

And they are miserable.

This is the danger of Door #2 (The Zoo).

If the government gives us a UBI (allowance) but denies us Agency (ownership/purpose), we will become Universe 25.

We will become a society of narcissists grooming ourselves in the mirror while the species slowly dies out.

The "Crisis of Meaning" is not a philosophical debate for smoke-filled rooms. It is a biological emergency.

If we remove the external pressure of survival, we must find a way to replace it, or our psychology will collapse.

Universal Truth: Resistance is not the enemy of life. Resistance IS life.

Muscles atrophy without gravity. Bones dissolve in space. The human spirit dissolves in paradise.

Section 4: The Physics of Meaning (Voluntary Struggle)

So, what is the solution? Do we destroy the machines and go

back to plowing fields with oxen just to feel "alive"?

No. That is the Luddite fallacy. We cannot un-invent the tractor.

The solution is to shift from **Involuntary Struggle** (Survival) to **Voluntary Struggle** (Mastery).

We must transition from "I suffer because I have to" to "I suffer because I choose to."

This is the concept of The Anti-Atrophy.

Just as an astronaut on the International Space Station must exercise for two hours a day on a specialized treadmill to prevent their body from dissolving, the citizen of the Dividend Era must exercise their will to prevent their soul from dissolving.

We must engineer our own gravity.

This brings us to the concept of Misogi.

In the Shinto tradition, Misogi is a ritual of purification through cold water. It is a voluntary shock to the system.

In the modern context, it means choosing a challenge so difficult that it defines you.

- **The Marathon:** Why do people run 26.2 miles? It costs money. It hurts. It ruins your knees. There is no economic benefit. They do it because the *pain validates their existence.*
- **Jiu-Jitsu:** Why do tech billionaires wake up at 6 AM to get strangled by a sweaty man in a gym? Because the simulation of death makes them feel alive.
- **Mountaineering:** Climbing Everest is the ultimate irrational act. It is expensive, useless, and deadly. Yet the waiting list is full.

In the Dividend Era, these "hobbies" will cease to be hobbies. They will become Religions.

They will become the primary structure of our lives.

We will not define ourselves by our "Job" (Accountant, Lawyer,

Driver). We will define ourselves by our "Struggle."

"I am a Climber."

"I am a Poet."

"I am a Builder."

The "Void" is filled by the friction of self-imposed difficulty.

Section 5: The Return of the Craftsman (Meaningful Inefficiency)

This leads to a paradoxical economic prediction: **The Boom of Inefficiency.**

In the era of AI, efficiency is free.

If you want a chair, a robot can print a perfect plastic chair for $1. It is symmetrical, durable, and cheap.

But you won't want that chair. You will want the wooden chair made by your neighbor, John.

John's chair is imperfect. One leg is slightly shorter. It squeaks. It took him 40 hours to carve it by hand. It costs $500.

You will buy John's chair.

Why?

Because John's chair contains Time. It contains Struggle. It contains Humanity.

As AI commoditizes "Perfection," humans will place a premium on "Flaws."

We will see a massive resurgence of the Craftsman Economy.

- Hand-grown tomatoes (inefficient).
- Hand-knitted sweaters (inefficient).
- Live theater (inefficient).
- Philosophy circles (inefficient).

We will pay for things *because* they were hard to make. We will pay for the "Proof of Life" embedded in the object.

This is the "Singularity of Value" we discussed in Part II, but now applied to the soul.

Work will no longer be about "Production" (the robots do that). Work will be about "Expression."

The future is not Wall-E (fat people in floating chairs).

The future is ancient Athens.

In Athens, the citizens didn't work in the fields (slaves did that). Did the Athenians become "Beautiful Ones"? No. They invented Geometry. They invented Democracy. They invented Tragedy. They spent their days in the gymnasium (training the body) and the agora (training the mind).

They used their freedom to pursue **Arete** (Excellence).

We are about to become a civilization of 8 Billion Athenians.

(Or 8 Billion Mice. The choice is ours).

Section 6: The Ubermensch

Friedrich Nietzsche predicted this crisis 150 years ago.

He famously said, "God is dead."

He wasn't celebrating atheism. He was warning us. He meant that the old structure of meaning (Religion/Tradition) had collapsed, and we were staring into the abyss of Nihilism.

He warned of the "Last Man" (The Beautiful One).

The Last Man is a creature who seeks only comfort. He blinks and says, "We have invented happiness." He has no great passion, no great star. He is a bug.

But Nietzsche also offered the solution: The Ubermensch (The Overman).

The Ubermensch is the one who creates his own values. He is the one who looks at the Void and builds a cathedral in it. He does not need a boss or a god to tell him what to do. He commands himself.

The Dividend Mandate is the platform for the Ubermensch.

It removes the excuse of necessity.

You can no longer say, "I would write my novel, but I have to work at the bank."

The bank is gone. The check is in the mail.

Write the novel.

If you don't write it now, it's not because you were busy. It's because you were empty.

This is the terrifying freedom of the Golden Age.

When the cage door opens, you have to decide to fly. The zoo keeper isn't going to push you.

The crisis of the 21st century will not be a crisis of poverty. It will be a crisis of **Will**.

Universal Truth: You must become the Architect of your own suffering, or the Void will consume you.

We have now covered the Diagnosis, the Theory, the Solution, the Defense, and the Meaning.

The book is complete.

We are ready to execute the final protocol.

SYSTEM LOG: SOUL DIAGNOSTIC

STATUS: EXISTENTIAL RISK IDENTIFIED

SUBJECT: THE VOID & THE MOUSE UTOPIA

SUMMARY:

- **Universe 25:** When struggle is removed, a population does not become happy; it becomes autistic and self-destructive. "The Beautiful Ones" die out.
- **The Crisis of Meaning:** Without the necessity of survival, humans face the "Void." We must engineer our own purpose.
- **Voluntary Struggle:** The antidote to atrophy is self-imposed hardship (Misogi). We must choose our mountains.
- **The Craftsman's Return:** As efficiency becomes free, inefficiency (art, philosophy, handmade goods) becomes the new luxury.

ACTION REQUIRED: PROCEED TO PART VIII FOR COGNITIVE UPGRADE.

PART VIII: THE EDUCATION REBOOT

CHAPTER 16: THE ACADEMY OF THE NEW MIND

Section 1: The Factory of the Mind

Walk into a modern public school. Look around. What do you see?

You see children sitting in straight rows. You see them moving from room to room at the sound of a mechanical bell. You see them raising their hands to ask permission to speak or use the bathroom. You see them memorizing facts to regurgitate them onto a standardized form.

Now, walk into a 19th-century textile factory.

You see men standing in straight rows. You see them moving at the sound of a shift bell. You see them asking the foreman for permission. You see them performing repetitive tasks.

The resemblance is not accidental. It is architectural.

Our current education system is not broken. It is working exactly as it was designed. The problem is that it was designed in **Prussia in the early 1800s**.

The Prussian Model was not created to produce thinkers, philosophers, or innovators. It was created to produce Obedient Soldiers and Compliant Factory Workers.

The state needed biological machines that could follow

orders, tolerate boredom, and execute repetitive tasks without questioning authority.

- **The Bell** conditions you to let an external authority dictate your time.
- **The Rows** condition you to isolation and surveillance.
- **The Grades** condition you to external validation.

For 200 years, this system worked. The economy *needed* obedient workers. It needed human calculators. It needed human filing cabinets.

But in 2026, this system is a crime against humanity.

We are training children to be robots in an age where robots are a billion times faster, cheaper, and more obedient than they are. We are training them to compete with AI on AI's home turf (memorization and compliance).

This is a losing battle.

If you train a child to be a "Bad Robot"—a creature that memorizes facts and follows rules—you are training them for the unemployment line.

We must tear down the Prussian Model. We must stop building factories for the mind and start building **Gymnasiums for the Soul**.

Section 2: The Death of the Report Card

The most dangerous metric in the world is the **GPA (Grade Point Average)**.

We tell our children: "Get straight A's. That means you are smart."

We are lying.

"Straight A's" does not measure Intelligence. It measures Compliance.

To get an A, you must:

1. Sit still.
2. Listen to the teacher.
3. Memorize the provided information.
4. Repeat it back exactly as requested.

Guess who gets straight A's effortlessly?

ChatGPT.

An AI can pass the Bar Exam, the Medical Boards, and the SATs in the 99th percentile. It has perfect recall. It never gets bored. It never questions the prompt.

By the metric of the Report Card, the AI is the perfect student.

If your child's value proposition is "I can answer the question correctly," they are obsolete. The answer is now free. The answer is a commodity.

We need to invert the metric.

We shouldn't grade children on the Answer. We should grade them on the Question.

- **The Old Metric:** "Who was the 16th President?" (Memory).
- **The New Metric:** "Why does the concept of the Presidency matter in a decentralized world?" (Synthesis).

We need to stop asking: "Did you follow the instructions?"

We need to start asking: "Did you break the instructions in a way that created new value?"

The "A Student" works for the "C Student" because the C Student was usually busy questioning the system while the A Student was busy obeying it. In the AI era, the "C Student"—the disruptor, the creative, the disagreeable—is the only one who survives.

Section 3: Stop Learning Syntax (The Coding Fallacy)

There is a panic among parents right now. "My kid needs to learn to code! Coding is the literacy of the future!"

This is the "Learn to Weld" of the 21st century. It is well-intentioned, but it is shortsighted.

Do not teach your children Syntax.

Syntax (Python, C++, Java) is the "vocabulary" of coding. It is the placement of semicolons and brackets.

AI is already better at Syntax than any human. GitHub Copilot writes the boilerplate code.

In five years, "coding" will simply be talking to a computer in natural English.

"Computer, build me a website that looks like Apple.com but sells shoes."

The AI handles the syntax.

Instead of Syntax, teach **Logic** and **Architecture**.

- **Logic:** How do you break a complex problem into solvable steps? (Algorithms).
- **Architecture:** How do the systems fit together? How does the database talk to the frontend?

Teach them Systems Thinking.

A great architect doesn't need to know how to mix the concrete by hand. They need to know how the building stands up against the wind.

If you teach a kid Python, you give them a job for 5 years.

If you teach a kid Logic, you give them a career for 50 years.

Universal Truth: Syntax changes every decade. Logic is eternal.

Section 4: The New 3 R's

The traditional "3 R's"—Reading, Writing, Arithmetic—are now the domain of the machine. The AI reads faster, writes cleaner, and calculates instantly.

We must pivot to the **New 3 R's** of the Human Age.

1. Rhetoric (Persuasion & Storytelling)

In a world of infinite data, the ability to move people is the ultimate leverage.

AI can generate a report. It cannot stand in a room, look people in the eye, and convince them to follow a vision.

We must teach Rhetoric. We must teach public speaking, negotiation, and storytelling.

The ability to capture attention and direct it is the superpower of the 21st century. If you can't sell your idea, the algorithm will bury it.

2. Research (Truth Verification)

We are entering the "Post-Truth" era. Deepfakes, hallucinations, and bot armies will flood reality with sludge.

The most valuable skill is no longer "finding information" (Google does that). It is "Vetting Information."

We must teach children to be Epistemological Detectives.

- "How do you know this is true?"
- "What is the source?"
- "What is the bias?"
- "Is this video real?"
    This is Intellectual Self-Defense.

3. Resilience (Mental Toughness)

This connects back to Chapter 12 (The Human Fortress).

The modern world is designed to make children fragile. It protects them from failure ("Participation Trophies") and bombards them with dopamine.

We must teach Stoicism. We must teach them how to sit with boredom. We must teach them how to fail publicly and recover.

A child who collapses at the first sign of adversity will be crushed by the "Great Turbulence." A child who views failure as data will thrive.

Section 5: The Cyborg Apprentice (Aristotle in the Pocket)

So, what does the school of the future look like?

It doesn't look like a school.

It looks like **The Young Lady's Illustrated Primer** from Neal Stephenson's *The Diamond Age*.

For the first time in history, we can solve "Bloom's 2 Sigma Problem."

Educational psychologist Benjamin Bloom proved that a student with a 1-on-1 tutor performs two standard deviations (98th percentile) better than a student in a classroom.

But we couldn't afford a human tutor for every child.

Now we can.

Every child will have an AI Tutor in their pocket.

- It knows their learning style.
- It knows their interests (if they like Minecraft, it teaches math using Minecraft blocks).
- It never gets tired.
- It never judges them for asking "stupid" questions.

This allows for Hyper-Personalized Unschooling.

The child doesn't need to sit in a room with 30 other kids moving at the same pace. The fast kid can fly. The slow kid can take their time to master the foundation.

The role of the human "Teacher" shifts.

The Teacher is no longer the "Sage on the Stage" (dispensing facts).

The Teacher becomes the "Guide on the Side" (Mentorship).

The human teacher focuses on the social, the emotional, and the philosophical. The AI handles the technical transfer of knowledge.

We are raising Cyborg Apprentices—humans who naturally integrate AI into their cognitive loop, treating the machine not as a cheating device, but as an exoskeleton for their mind.

Section 6: The Return of the Polymath

The Industrial Age loved Specialization.

"Learn one thing. Do it for 40 years. Retire."

This is the strategy of an insect. It is efficient for a stable environment.

But the AI environment is chaotic. The Specialist is vulnerable. If AI learns your "One Thing," you are dead.

The future belongs to the Generalist. The Polymath.

The person who can connect dots between disparate fields.

- The Biologist who understands Economics.
- The Coder who understands Philosophy.
- The Artist who understands Physics.

Innovation happens at the intersections. AI is bad at intersections (it is trained on specific datasets). Humans are great at metaphor and cross-pollination.

We need to encourage Intellectual Range.

Let the child learn the violin AND coding AND gardening.

Do not force them to pick a "Major" too early.

The wider their base, the higher their pyramid can go.

SYSTEM LOG: COGNITIVE UPGRADE

STATUS: RE-EDUCATION INITIATED

SUBJECT: THE ACADEMY OF THE NEW MIND

SUMMARY:

- **The Prussian Failure:** The current school system trains children to be obedient factory workers. This is obsolete. We are training "Bad Robots."
- **Death of the GPA:** Grades measure compliance, not intelligence. AI gets straight A's effortlessly. We must measure curiosity and synthesis.
- **The New 3 R's:** Replace Reading/Writing/Arithmetic with Rhetoric (Persuasion), Research (Truth), and Resilience (Grit).
- **The Polymath:** Specialization is for insects. The future belongs to Generalists who can connect dots across disciplines.

ACTION REQUIRED: PROCEED TO PART IX FOR BIOLOGICAL DEPLOYMENT.

PART IX: THE BLUEPRINT 2.0 (BIOLOGY & LAW)

CHAPTER 17: THE METHUSELAH PROJECT (BIOLOGICAL EQUITY)

Section 1: The Final Inequality

We have talked about money. We have talked about power. We have talked about meaning. But there is one asset that makes all others irrelevant.

Time.

If you have a billion dollars in the bank, but you have Stage 4 Pancreatic Cancer, you are poor. If you have $0 in your pocket, but you are 20 years old and healthy, you are rich.

For all of human history, Death was the great equalizer. The King and the Pauper both aged. Both got sick. Both died around 70 or 80. Biology did not discriminate based on your zip code.

That is about to change. We are standing on the precipice of **Biological Escape Velocity**.

Artificial Intelligence has cracked the code of life. In 2020, DeepMind's AlphaFold solved the "Protein Folding Problem"—a grand challenge of biology that had stumped human scientists for 50 years. In a few weeks, the AI predicted the 3D shape of nearly every protein known to science.

This turned Biology from a mystery into an engineering discipline. We are no longer "discovering" drugs; we are **designing** them.

The Elites know this. Look where the smart money is going.

- Jeff Bezos invested billions in **Altos Labs** (cellular rejuvenation reprogramming).
- Sam Altman invested $180 million into **Retro Biosciences** (adding 10 years to healthy lifespan).
- The Saudis are funding the **Hevolution Foundation** ($1 billion/year) to treat aging as a disease.

They are not trying to cure the flu. They are trying to cure Death.

And they are going to succeed.

But here is the terrifying question: **Who gets the cure?**

In a profit-maximalist system, longevity treatments will be the ultimate luxury good.

Imagine a gene therapy that reverses aging by 20 years. It costs $5 million per dose.

- The Elite live to be 150, remaining physically young, accumulating compound interest, and consolidating power for nearly two centuries.
- You? You age. You break down. You serve them. You die at 75.

This creates a **Biological Caste System**. It creates a literal divergence of the species: *Homo Deus* (The Immortal Rich) and *Homo Sapiens* (The Mortal Poor).

This is the ultimate nightmare of the Zero Economy.

Section 2: The Right to Repair (Your Own Body)

The Dividend Mandate must extend beyond the wallet. It must enter the veins. We must demand **Biological Equity**.

The argument is simple: The AI that discovered the cure was

trained on Human Data.

It was trained on the collective biological knowledge of our species—our clinical trials, our genomes, our medical history.

Therefore, the fruit of that AI belongs to the species, not just the shareholders of the biotech firm.

We must categorize Longevity Technology not as a "Luxury Good" (like a Ferrari), but as a **"Public Utility"** (like a Vaccine or Clean Water).

This is The Methuselah Mandate.

It states: Any life-extension therapy developed using Sovereign AI models must be made available to every citizen at the marginal cost of production.

If the pill costs $0.50 to manufacture, it sells for $0.50. We cannot allow the "Patent Wall" to fence off the fountain of youth.

This is not just moral; it is economic.

An aging population is a bankrupt population. Alzheimer's and heart disease cost the economy trillions in lost productivity and care.

A population that is biologically 25 years old for 100 years is a population of Super-Producers. They are robust. They are anti-fragile. They do not burden the state; they build the state.

The "Methuselah Project" is the plan to use the Sovereign Fund to subsidize the mass deployment of rejuvenation biotechnology. We treat Aging as a pandemic. And we treat the cure as a human right.

Section 3: The Data Donor (The Biological Swap)

How do we pay for this? We pay with our bodies.

Currently, you give your medical data away for free, or it is siloed in hospitals that don't talk to each other. It is wasted.

Under the Dividend Mandate, we create a National Bio-Bank.

This is the "Grand Swap" applied to biology.

The Deal:

- **You Opt-In:** You agree to share your anonymized genome, your blood markers, and your real-time biometric data (from your wearable devices) with the Sovereign AI.
- **You Receive:** Free access to the longevity treatments derived from that data.

You become a "Data Donor."

The AI uses your body to learn how to cure disease. You get the cure as the dividend.

It is a perfect feedback loop. The more citizens join, the smarter the AI gets. The smarter the AI gets, the faster the cures arrive.

The Elites will try to hoard the tech. They will want to be the gods on Olympus, looking down on the mortals.

We must remind them: Olympus was built on our backs. And we intend to live there too.

CHAPTER 18: THE CODE OF LAW (THE BLUEPRINT)

Section 1: The Paper Shield

We have the ideas. Now we need the armor.

If we pass the Dividend Mandate as a simple "law" or "policy," we will lose. The politicians will repeal it the moment we look away. They will loot the Sovereign Fund to pay for wars or bailouts. They will dilute the currency.

We cannot trust politicians. We can only trust **Code**.

We need to upgrade the Operating System of the United States Constitution. We need three specific mechanisms to lock this revolution in stone, making it tamper-proof against future corruption.

Section 2: The 28th Amendment (The Data Rights)

The US Constitution protects your speech (1st), your guns (2nd), and your property (4th/5th).

But it does not protect your Data.

When the Constitution was written, "Data" didn't exist. Ben Franklin couldn't imagine a Cookie, a Neural Net, or a predictive algorithm.

Because of this loophole, the Courts have ruled that you have no "expectation of privacy" for data you give to a third party (The

Third-Party Doctrine).

We need a **28th Amendment**.

Draft Text:

"The right of the people to be secure in their digital persons, biometric data, and cognitive outputs against unreasonable search, seizure, or commercial extraction without explicit consent and compensation, shall not be violated."

This changes everything.

- It makes "Data Mining without Payment" unconstitutional.
- It forces Facebook, Google, and OpenAI to sit at the negotiating table.
- It turns "Data Serfdom" into "Data Property Rights."

Without this Amendment, we are fighting a war with no legal standing. With it, we have the Supreme Court. We transform our digital footprint from "exhaust fumes" into "gold."

Section 3: The DAO of America (Trustless Treasury)

Where do we put the money?

If we put the Sovereign Fund in the hands of the US Treasury or the Federal Reserve, it will be corrupted. It will be "misallocated" to special interests.

We need to move the Treasury to the Blockchain.

We need to create the American DAO (Decentralized Autonomous Organization).

This is not sci-fi. It is simply applying transparency technology to public finance.

1. **Transparency:** Every dollar that enters the Fund (from the Compute Tax or Equity Swap) is visible on a public ledger. You can verify it on your phone. No black budgets. No missing trillions.
2. **Immutability:** The distribution rules are written in "Smart

Contracts."

- ○ *Code Rule:* "If Fund Balance > X, Then distribute Y to all Citizen Wallets."
- ○ No politician can intervene. No committee can "pause" it to fund a bridge to nowhere.

3. **Direct Voting:** Shareholders (Citizens) can vote on high-level allocation decisions using their cryptographically secured identity.

We replace "Trust in Bureaucrats" with "Trust in Math."

The "Check" doesn't come from the President. It comes from the Protocol.

Section 4: The Exit Tax (The Golden Handcuffs)

The billionaires will threaten to leave.

They will say: "If you tax our compute, if you force us to give equity, we will move to Dubai! We will move to Mars!"

Let them go. But they cannot take the Assets.

We implement a draconian **IP Exit Tax**.

The logic is simple: If a company developed its AI using American Data, American Energy, American Universities, and American Protections, that Intellectual Property is domiciled *here*.

If they want to re-incorporate in the Cayman Islands to avoid the Dividend Mandate, they must pay an Exit Tax equal to **100% of the value of the Intellectual Property**.

Effectively: **You can leave, but you leave naked.**

You cannot strip-mine the American people and then run away with the gold. This creates "Golden Handcuffs." It forces the Elites to stay and play the game (Door #3). It forces them to accept the Dividend as the cost of doing business in the most lucid, stable, and wealthy market on earth.

Section 5: The Final Assembly

We have the Medical Tech (Chapter 17).

We have the Legal Code (Chapter 18).

The Blueprint is complete. The machine is designed.

All that remains is the destination.

Where does this ship actually go once we leave the harbor? What does the "Golden Age" actually feel like?

We are ready for the final chapter.

SYSTEM LOG: BLUEPRINT FINALIZED

STATUS: READY FOR DEPLOYMENT

SUBJECT: BIOLOGY & LAW

SUMMARY:

- **The Methuselah Mandate:** Longevity technology must be a public utility, not a luxury good. We demand biological equity.
- **The Data Donor:** We trade our biological data for the cure to aging. It is the ultimate Grand Swap.
- **The 28th Amendment:** We need constitutional protection for our digital selves. Data property rights must be enshrined in law.
- **The American DAO:** The Treasury must move to the blockchain. Trust in math replaces trust in bureaucrats.

ACTION REQUIRED: PROCEED TO PART X FOR FINAL ACTIVATION.

PART X: THE DESTINATION

CHAPTER 19: THE GOLDEN AGE (MANIFESTO)

Section 1: The Morning After the War

Tuesday, October 14, 2036.

Open your eyes. Listen.

Do you hear it?

No. You don't.

There is no grinding of metal. There is no screaming foreman. There is no blaring alarm clock shattering your dreams at 6:00 AM. There is no rush hour traffic humming on the highway like a dying beast.

There is just the wind in the trees and the sound of your children laughing in the kitchen.

You wake up because the sun hit your face, not because a corporation demanded your presence. You stretch. You are not tired. You are not anxious. The tightness in your chest—the low-level panic about rent, bills, and layoffs that defined the 2020s—is gone.

You pick up your device. It is a thin sheet of glass, indistinguishable from magic. You check your notifications.

There is no email from a boss. There is no Slack message demanding an update.

There is just one notification from the American DAO.

"Q3 DIVIDEND DEPOSITED: $4,500."

You smile. You swipe it away. You don't frantically calculate if this covers the mortgage. You know it does. You know the floor is solid.

You walk to the kitchen. The coffee is ready. The house is clean. The household bots—silent, efficient, invisible—have done the laundry, scrubbed the floors, and ordered the groceries.

The "drudgery" of existence has been automated. The "struggle" of survival has been solved.

You walk out onto the porch. You look at your town. It looks different than it did ten years ago. The strip malls are gone. The fast-food chains are gone.

In their place are parks, workshops, art studios, and community gardens.

The "Job" is dead. But "Work" is alive.

You see your neighbor, Sarah. She used to be an exhausted data entry clerk, terrified of being replaced by a script. Now, she runs the local theater company. She is rehearsing a play. She looks ten years younger.

You see your other neighbor, Mike. He used to drive a truck for 14 hours a day, his back destroying itself on the seat. Now, the trucks drive themselves. Mike spends his days teaching history to the neighborhood kids and restoring vintage motorcycles.

They are not "unemployed." They are Liberated.

This is the world we fought for. This is the Dividend Era.

Section 2: The End of Time Slavery

To understand the magnitude of this victory, we must look back at the enemy we defeated.

We didn't just defeat poverty. We defeated **Time Slavery**.

For 300 years, since the first steam engine hissed to life, the human experience was defined by the commodification of time.

- You sold your hours to buy your life.
- You traded 8 hours of boredom for 4 hours of freedom.
- You missed your child's first steps because you were in a Zoom meeting.
- You missed the sunset because you were stuck in traffic.

We accepted this as "natural." We thought it was the price of civilization.

It wasn't. It was a temporary bug in the operating system of history. It was a bridging period where machines were dumb and humans had to act like robots.

But in 2036, the bridge is built.

The AI does the thinking. The robots do the lifting. The fusion reactors provide the power.

- The "Cost of Living" has collapsed to near zero (The Deflationary Black Hole).
- The "Price of Goods" has collapsed to near zero.
- And because we secured the Dividend, the "Access to Wealth" is universal.

You have reclaimed your Time.

What do you do with it?

The cynics of the 2020s (The "Counter-Reformation" crowd) said that if you gave people money without work, they would become lazy. They said we would all sit on the couch, drugged by VR, eating Cheetos. They pointed to the "Mouse Utopia" and warned of collapse.

They were wrong.

They confused "Exhaustion" with "Laziness."

People were lazy in 2024 because they were burned out. They

were fried.

In 2036, with the boot off our necks, the human spirit has exploded. We are seeing a **Renaissance of the Real**.

People are not doing *nothing*. They are doing the things that machines *cannot* do.

- We are exploring the biology of our own planet.
- We are creating art that is messy and raw and undeniably human.
- We are debating philosophy in the public square.
- We are raising our children with a presence that no generation has ever had.
- We are building communities based on shared interest, not shared commute.

The "Hamster Wheel" has stopped spinning. The cage door is open. And we have stepped out into the sun.

Section 3: The New Aristocracy

In ancient Greece, there was a class of people who did not work for wages. They were the Aristocracy. Because they didn't have to toil in the fields, they invented Geometry. They invented Democracy. They invented Tragedy.

But their freedom was built on a crime: Slavery. They relied on human slaves to do the work.

In 2036, we are **all** Aristocrats.

We have slaves, but they are made of silicon and steel. We have millions of them. They mine the lithium. They harvest the wheat. They code the software. They clean the sewers.

They do not suffer. They do not complain. They simply serve.

We have democratized the privilege of the leisure class. We have taken the lifestyle of the 19th-century gentleman—the time to think, read, and tinker—and given it to the truck driver and the waitress.

This is the **Human Premium**.

In this new world, status is not determined by your Job Title. Nobody asks, "What do you do?" at parties anymore. That question is obsolete.

They ask: "What are you building?"

They ask: "Who are you loving?"

They ask: "What are you learning?"

Status is derived from your contribution to the human layer of reality.

The billionaire is not the guy with the most money (money is abundant). The billionaire is the guy with the most interesting dinner table.

We have moved from an economy of Transaction to an economy of Relation.

Section 4: The Ghost of the Machine

Go to the edge of town. Look at the factory.

It is a massive, black monolith. It hums with a low, vibrating power. There are no windows. There are no parking lots.

Inside, the lights are off. The machines don't need light to see.

Inside, the robots are moving at speeds the human eye can't track. They are assembling, welding, printing, and packaging.

Inside, the AI is running a trillion simulations a second to optimize the supply chain.

It is terrifyingly efficient. It is alien.

But look at the sign on the gate. It doesn't say "Property of Amazon." It doesn't say "Property of BlackRock."

It says: **"PROPERTY OF THE AMERICAN SOVEREIGN FUND."**

You own that factory.

You own the robots in the dark.

Every time a robot arm swings, a fraction of a cent drops into your digital wallet.

Every time the AI solves a problem, your life gets a little richer.

We didn't smash the machines (Luddites).

We didn't let the machines enslave us (The Matrix).

We bought the machines.

We harnessed the power of the Singularity and put a bridle on it. We turned the monster into a beast of burden.

This is the victory of **Dividendism**.

Section 5: The Final Warning

But snap back to reality.

Look at the date on your phone. It is not 2036. It is **2026**.

You are not on the porch. You are in the trenches.

The factory is being built, but you don't own it yet. The AI is being trained, but it is being trained to replace you, not serve you. The laws are not passed. The Fund does not exist.

The vision I just described is not a promise. It is a **Possibility**.

It is one of the Three Doors. And right now, the wind is blowing it shut.

- The forces of Door #1 (The Bunker) are strong. They are hoarding the chips. They are buying the politicians. They are building the walls.
- The forces of Door #2 (The Zoo) are loud. They are printing the money. They are preparing the sedative.

If you do nothing—if you put this book down and go back to scrolling TikTok—we will lose.

We will slide into the abyss of Techno-Feudalism. You will

become a serf. Your children will be pets. The Golden Age will be stolen from you, just like the Commons were stolen in 1773.

The next five years determine the next five hundred.

This is the Checkpoint.

Section 6: The Architect's Charge

I have given you the map.

- **You have the Diagnosis:** You know the Wage System is dead. You know the engine is broken.
- **You have the Theory:** You understand the Physics of Zero. You know why deflation is inevitable.
- **You have the Solution:** You know about the Sovereign Fund, the Compute Tax, and the Grand Swap.
- **You have the Action Plan:** You know how to build the Barbell. You know how to build the Human Fortress.

The intellectual work is done. Now comes the heavy lifting.

1. **Secure Your Sovereignty:** Get out of debt. Get out of cash. Buy the Shield (Gold/Land). Buy the Spear (AI Equity). Harden your mind. Unplug from the Matrix.
2. **Spread the Mandate:** Ideas are viruses. Infect your network. Talk about the Dividend at the dinner table. Talk about the Sovereign Fund at the bar. Stop complaining about "Jobs." Start demanding "Equity." Shift the Overton Window.
3. **Hold the Line:** When the "Great Turbulence" hits —when the layoffs start, when the banks wobble, when the fear spreads—do not panic. Remember the destination. Remember the porch in 2036.

The chaos is necessary. It is the birth pangs of the new world.

We are the generation that stands between the dark and the light.

We are the last generation of workers and the first generation of

165

free humans.

It is a heavy burden. But it is a glorious burden.

Do not let them scare you.

Do not let them buy you off with cheap distractions.

Do not let them tell you that you are "obsolete."

You are the Shareholder. You are the Master. You are the Point of the Spear.

The robots are here to work.

We are here to live.

The blueprints are in your hands.

Build the future.

EPILOGUE: THE WAGER (LETTER TO 2050)

To the Child born today:

I am writing this from the year 2026. It is a scary time to be alive. The world feels like it is breaking. The old maps are burning, and we don't have new ones yet. People are angry. People are confused.

We are staring at a new intelligence—an alien mind made of silicon—and we are wondering if it is our savior or our executioner.

I don't know if we won.

I don't know if you are reading this from a Golden Age, where you own your time, your biology is optimized, and the machines serve you as loyal stewards.

Or if you are reading this from a digital cage, scanning this text in secret, wondering what the word "Freedom" actually meant.

But I want you to know that **we tried**.

We saw the train coming. We saw the math. We stood in the server rooms and felt the heat of the future rising. We tried to grab the steering wheel before the engine went off the cliff.

We wrote this book not because we were certain of victory, but because we refused to accept defeat.

We placed a **Wager on Humanity**.

- We bet that the human spirit was stronger than the algorithm.
- We bet that our desire for meaning was stronger than their desire for control.
- We bet that we could build a system where technology amplifies freedom instead of extinguishing it.

If you are reading this, and you are free... then the wager paid off.

Do not take it for granted. The jungle always tries to grow back over the garden. The tyrants always try to rebuild the walls. The code always tries to rot.

Guard the Dividend.

Guard the Truth.

Guard your Sovereignty.

The light is yours now. Keep it burning.

— The Architect

SYSTEM REPOSITORY: THE SOURCE CODE

This manual was not written in a vacuum. It was constructed by reverse-engineering the economic, historical, and biological data of the last three centuries.

For those who wish to inspect the raw code, these are the primary libraries.

I. THE KERNEL (ECONOMIC THEORY)

- **Rifkin, Jeremy.** *The Zero Marginal Cost Society.* (The physics of the price collapse).
- **Paine, Thomas.** *Agrarian Justice.* (The original blueprint for the Citizen's Dividend).
- **Kurzweil, Ray.** *The Singularity Is Near.* (The trajectory of exponential tech).
- **Keynes, John Maynard.** *Economic Possibilities for our Grandchildren.* (The prediction of technological unemployment).

II. THE FIREWALL (GEOPOLITICS & HISTORY)

- **Allison, Graham.** *Destined for War: Can America and China Escape Thucydides's Trap?* (The logic of the AI Arms Race).
- **Miller, Chris.** *Chip War: The Fight for the World's Most Critical Technology.* (The battle for Taiwan).
- **Sale, Kirkpatrick.** *Rebels Against the Future.* (The true history of the Luddites).

III. THE BIOS (HUMANITY & BIOLOGY)

- **Sinclair, David.** *Lifespan: Why We Age—and Why We Don't Have To.* (The science of the Methuselah Project).
- **Calhoun, John B.** *Death Squared: The Explosive Growth and Demise of a Mouse Population.* (The warning of Universe 25).
- **Frankl, Viktor.** *Man's Search for Meaning.* (The antidote to the Void).

IV. THE UI (EDUCATION & FOCUS)

- **Stephenson, Neal.** *The Diamond Age.* (The blueprint for the AI Tutor).
- **Newport, Cal.** *Deep Work.* (The protocol for the Human Fortress).
- **Taleb, Nassim Nicholas.** *Antifragile.* (The strategy for surviving chaos).

ABOUT THE ARCHITECT

DAVID KHOA NGUYEN

System Architect. Engineer. Macro-Strategist.

David does not view the world through the lens of politics or ideology. He views it through the lens of **Physics** and **Code**.

Operating at the intersection of Hardware and Economics, David serves as a Test & Debug Architect for next-generation computing infrastructure. He works daily with the silicon engines—the NVIDIA H200s and Blackwell B300 superclusters—that are currently rewriting the operating system of the global economy.

He sees the "Zero Point" approaching not as a theoretical concept, but as a tangible reality in the heat and voltage of the server room.

He realized that while our **Hardware** (AI) is entering the 22nd Century, our **Software** (The Wage System) is stuck in 1913. This incompatibility is the source of the world's chaos.

He wrote *DIVIDENDISM* to provide the Patch.

David is not a politician. He is an engineer who believes that poverty is not a moral failing—it is a design flaw. And like any flaw in a system, it can be fixed with the right code.

Connect with the Architect:

- **System Access:** www.nguyenkdavid.com

- **Direct Line:** architect@nguyenkdavid.com

SYSTEM VERIFICATION: THE PROPAGATION PROTOCOL

The update is complete. You now hold the blueprints for the Post-Labor Age.

But code that sits on a hard drive is useless. Code must run. Code must spread.

We are fighting a war against **"Legacy Thinking."** The media, the politicians, and the institutions will try to convince you that the old rules still apply. They will gaslight you into believing that your poverty is your fault, not a system error.

We need more Architects.

If this book has re-wired your understanding of the economy —if you now see the "Glitch" in the wage system—you have an obligation to signal the network.

EXECUTE THE PROTOCOL:

1. Verify the Install:

Leave an honest review. The algorithm suppresses silence. It rewards noise. Make noise. Let the system know that the users are waking up.

2. Share the Key:

Hand this book to someone who is "asleep"—a friend worried about their job, a colleague trapped in the rat race, or a parent confused by the changing world. Wake them up.

The Dividend Era does not begin with a law passed in Congress.

It begins with a shift in your mind.

David Khoa Nguyen

The System Architect

2026